# Old Westminster Bridge

Nothing being more reasonable than that the Vices of private Persons should contribute as much as possible to the advantage of the Public; the Game of *Passage* is to be suppressed this year, as that of *Hazard* was the Last; the consequence of which, tis hoped will be, that *Annual State Lotteries* will from henceforth glean up all the Money which used to be confounded at Dice; and in due time furnish Posterity with a Monument at Westminster that may be called THE BRIDGE OF FOOLS through all Generations.

Henry Fielding

# OLD WESTMINSTER BRIDGE
## THE BRIDGE OF FOOLS

## RJB Walker

**David and Charles**
Newton Abbot   London   North Pomfret (Vt)

British Library Cataloguing in Publication Data
Walker, Richard John Boileau
 Old Westminster Bridge.
 1. London—Bridges—Westminster Bridge
 I. Title
 624.6'3'0942132     DA689.B8
 ISBN 0-7153-7837-6

Library of Congress Catalog Card Number 79-52379

Typeset by ABM Typographics Limited, Hull
Printed in Great Britain by
Redwood Burn Limited, Trowbridge and Esher
for David & Charles (Publishers) Limited
Brunel House   Newton Abbot   Devon

Published in the United States of America
by David & Charles Inc.
North Pomfret   Vermont 05053   USA

# Contents

# List of Plates

# 1
# The Need for a Bridge

After the Restoration the citizens of Westminster congratulated King
Charles on his return to his country. They admired the stupendous actings
of Divine Providence and gently reminded their sovereign that his honour
and safety were closely concerned in the good government of Westminster,
the place of his nativity and constant residence. Charles II courteously
agreed with his subjects but left it to them to implement their suggestion.
A series of Acts of Parliament was the result, aimed at improving the
Cities of London and Westminster by repairing the roads and sewers,
paving the streets and scavenging them every week, licensing the hackney
coaches, and trying to lighten the darkness by hanging out candles every
night from Michaelmas to Lady Day.[1]

The population increased relentlessly. The Plague is said to have killed
100,000 Londoners and about 3,000 in the parish of St Margaret's alone—
a large proportion considering that in spite of rapid building there were
then only some 1,400 houses in Westminster. One probable explanation
was the influx of scrofulous patients hoping to be touched for the King's
Evil; 200 were treated every Friday, amounting to 24,000 people between
the Restoration and the Plague. By 1682 Morden and Lea's map of London
shows the houses of Westminster extending up-river as far as the Horseferry, with the fields and orchards of Ebury Farm and the marshes of
Pimlico stretching to the villages of Chelsea and Knightsbridge.[2] Building
operations in St James's had begun soon after the Restoration, Pepys
noting in his diary for 2 September 1663, 'the building of St James's by
my Lord St Albans which is now about and which the City stomach, I
perceive, lightly but dare not oppose it'. The move westwards continued
along the Road to Knightsbridge (now called Piccadilly), Lord Clarendon
enclosing eight acres of ground to build Clarendon House immediately
after receiving the king's grant in 1664.[3] By 1721, Bond Street, begun in
1686 by Sir Thomas Bond 'banker and mechanic . . . as a street of tene-
ments to his undoing', had reached as far north as Oxford Street, the Road
to Tyburn. Berkeley House, Devonshire House and Burlington House

completed the great aristocratic palaces to the north of Piccadilly and set their seal on the general exodus of the nobility from the City and the banks of the Thames to the higher healthier ground to the west, leaving the Strand, Soho and Covent Garden to deteriorate into the rookeries of the eighteenth century, familiar to us from the terrifying engravings of Hogarth. William III and Queen Anne preferred to go even further west, to Hampton Court and Kensington Palace, returning to Whitehall and Westminster only for affairs of state and as infrequently as possible.

The growth of London had alarmed its citizens since the Tudors and throughout the seventeenth and early eighteenth centuries proclamations and Acts of Parliament were issued at intervals, restraining building activities and trying to control the proliferation of the slums, particularly in Westminster.[4] In January 1674, for instance, in a debate in the House of Commons, Sir William Coventry, on one of his rare visits to London from Minster Lovell, said that he would not like a 'beauty and uniformity in the City and a deformity at the King's court'. In the same debate Sir John Maynard said that this enlarging of London fills it with lacqueys and pages; and Sir John Duncombe, MP for Bury St Edmunds, said the growth of building westwards was becoming a positive danger, 'whole fields go into buildings and are turned into alehouses filled with necessitous people . . .[5] In fact long before Tucker had coined his celebrated phrase, 'a wen in the body politic', London had become easily the largest city in the world. By 1700 its population was in the region of 600,000, whereas in Paris the population was under 500,000 (in a much less densely populated country) and in Naples and Lisbon even less. Edinburgh, the next largest town in Britain, had 35,000, and Bristol and Norwich 20,000 each.

Defoe bemoaned its growth in that gold-mine of information about early eighteenth-century England, *A Tour through the Whole Island of Great Britain*. 'Whither will this monstrous City extend?' he asks at the end of a lamentation on selfish and haphazard building. Defoe's alarm reflected the concern of many people and the succession of Building Acts from the Fire to 1774 aimed at controlling the size and shape of houses but was met with complaints from the building trade. For instance a petition from a group of bricklayers, joiners, painters, carpenters, glaziers, brick-makers, tile-makers and many others, was presented to the House of Commons in 1710 protesting at a Bill intended to prevent building on new foundations which 'if passed into Law will be the utter Ruin of them and their numerous Families'.[6] But in order to avoid the worst horrors of poverty and squalor engendered by ramshackle dwellings thrown up in backyards and alley ways and on waste land, it was vital to exercise some sort of control. The picture of Georgian London consisting of elegant houses and terraced rows set in spacious squares filled with grass and trees is all too easily balanced by the horrific descriptions of town life by Fielding, Smollett, Dr Johnson, Hogarth and Gay. When the Bridge Commissioners were acquiring land in Westminster for the approaches to the new bridge, Thomas Lediard, the surveyor and agent responsible for buying property, reported almost continuously on the broken and tumble-down state of houses in the vicinity and the exorbitant prices their owners were demanding. Lediard reports on one occasion that 'great complaints' have been made to him and

> that some old houses in St Margaret's Lane within the limits put into the power of this Honble Commission by the last Act of Parliament, are in such a crazy condition and in so great danger of falling, that it would by no means be safe for His Majesty to go that way to the House of Lords till some care is taken of them. I have therefore taken another Surveyor to my assistance and viewed them narrowly, when to our great surprise, we found them so circumstanced that it is next to a miracle they have stood so long.[7]

If the houses in London were squalid and ramshackle, the streets were just as bad. Most of them were narrow, filthy, dark, ill-paved if paved at all, infested with thieves and dangerous to walk through after nightfall. César de Saussure, a young French observer travelling through England in 1725, does his best to be polite about the country but the streets defeat him. 'Do not expect me to describe all the streets of London', he writes home to his parents:

> I should have much to do and we should get tired of one another. A number of them are dirty, narrow and badly built; others again are wide and straight, bordered with fine houses. Most of the streets are wonderfully well-lit, for in front of each house hangs a lantern or a large globe of glass inside of which is a lamp burning all night. Large houses have two of these lamps suspended outside their doors by iron supports, and some have even four. The streets of London are unpleasantly full of either dust or mud. This arises from the quantity of houses that are continually being built, and also from the large number of coaches and chariots rolling in the streets day and night. Carts are used for removing mud, and in the summer the streets are watered by carts carrying barrels or casks pierced with holes through which the water flows. Another of the unpleasantnesses of the streets is that the pavement is so bad and rough that when you drive in a coach you are most cruelly shaken, whereas if you go on foot you have a nice smooth path paved with wide flat stones, and elevated above the road . . .[8]

13

De Saussure's descriptions tend to be optimistic and can be contrasted without difficulty from contemporary sources. Defoe, for instance, described King Street, the main street in Westminster, as 'a long, dark, dirty and very inconvenient Passage'.[9] In 1729 the inhabitants of Westminster presented a petition to the House of Commons complaining of 'the Filth and Dirt to the great Prejudice, Danger and Inconveniencing of all who inhabit or pass through the Streets of Westminster' and a committee was appointed to examine the problem. Christian Cole, a JP for Middlesex, said that 'the streets are in a very dirty and ruinous condition for want of being better paved and cleansed . . . the several kennels[10] are deep and dangerous and cannot be altered but by paving a whole street at once which cannot be done by the Laws in being'. He goes on to describe the confusion of authorities which were regulated by a combination of the Crown Office, the Justices of the Peace and the Borough Court of Westminster. The JPs themselves had gone in a body to view various streets and had fined the inhabitants, but quite ineffectually. It was further said that much of the bad condition of the streets was due to the Proprietors of the Waterworks laying down new pipes too near the surface and then paving with bad material (rubbish instead of gravel) on the wet ground over oozing pipes. In some streets the occupants had been forced to hire men to clean in front of their houses in addition to the regular scavengers.[11] The petition resulted in a Bill presented by the MP for Westminster, William Clayton, receiving the Royal Assent on 14 May 1729.[12]

The new Act gave more power to the Justices at special Sessions but the responsibility still lay with the householders and there was very little improvement, most occupants shirking their duties at the slightest opportunity. In 1739 another petition was presented to the Westminster Bridge Commissioners by the principal inhabitants and freeholders of the parish of St John the Evangelist compaining of the narrowness of Lindsay Lane, part of the area to be acquired for the opening of Old Palace Yard. 'Since the passing of the Act, some of the houses are become untenanted and others likely to be so or inhabited only by those who contribute but little towards enlightening or repairing the Publick Ways'[13]—a delicate stroke of understatement to indicate the motley horde of squatters occupying the houses due for demolition to make way for the bridge approaches.

By the end of the year the situation was getting out of control, complaints were made to the House of Commons and the Speaker wrote a letter to the Vestry of St Margaret's demanding action.

Gentlemen, Complaint having this day been made to the House of Commons that great Numbers of Idle and disorderly Persons do daily under pretence of asking Charity, infest the Streets and publick Places of the City and Liberty of Westminster to the great Annoyance and Interruption of the Members of the House of Commons in their Passage to and from the said House, and of other Persons going about their lawfull Occasions—notwithstanding the great Collections which have been made for the Relief of the Poor during the present severe season over and above the Provision of the respective Parishes for that purpose. The House of Commons has this day ordered me to send to the Justices of the Peace for the said City and Liberty and to the Vestries of the respective Parishes there, that Directions may be forthwith given by them to the Bedels, Constables and other Officers of the said Parishes, to put the Law in execution against all Idle and Disorderly Persons who shall be found infesting the streets and other publick places of the said City and Liberty under pretence of asking Charity, and in Obedience to the said Order I do hereby communicate the same to you.

I am, Gentlemen,

Your humble servant,

Ar. Onslow, Speaker

8th February 1739/40

As a result of this letter the Parish ordered the beadles to arrest 'all such Idle Persons as they shall find wandring and begging there, or using any subtle Craft or unlawful Game or Play, and particularly such as Game at Wheel Barrows ...[14]

However by 1741, when Westminster Bridge had been building for two years, the streets round its approaches were still inadequate and disgusting. Lord Tyrconnel raised the matter in the House of Commons and lives immortally through the words of Dr Johnson.

It is impossible, Sir, to come to this House, or to return from it, without observations on the present condition of the Streets of Westminster; observations forced upon every man, however inattentive or however engrossed by reflections of a different kind ... The filth, Sir, of some parts of the Town, and the inequality and ruggedness of others, cannot but in the eyes of foreigners disgrace our nation, and incline them to imagine us a people not only without delicacy, but without government—a herd of Barbarians or a colony of Hottentots. The most disgusting part of the character given by travellers of the most savage nations, is their neglect of cleanliness, of which perhaps no part of the world affords more proof than the streets of the British capital; a city famous for wealth, commerce and plenty, and for every other kind of civility and politeness, but which abounds with such heaps of filth as a savage would look on with amazement ...

Tyrconnel was supported by Lord Gage but snubbed by Samuel Sandys and his motion was rejected by 144 votes to 118[15]. It was not until long after the bridge had been finished that the Westminster Paving Act was passed in 1762 and steps were taken seriously to improve the condition of the streets.

As Dr Johnson said, London was a city famous for wealth, commerce and plenty. It was by far the largest and most important centre in the country from every point of view, social, economic, intellectual and cultural. It was the seat of government and law and it was swiftly approaching that position of world dominance that it was to enjoy for the next two hundred years. England in the first half of the eighteenth century was still, broadly speaking, an agricultural country dotted with towns, many of them increasing in size but all centring on the metropolis to which they dispatched their produce—food, drink and timber from most places, coal and glass from the Tyne, cloth from Lancashire, gloves from Worcester and Hereford. By the reign of George II, London had developed into an enormous market, drawing in produce from hundreds of miles around, absorbing a great deal into its own constitution and regurgitating a great deal more to the seaports and foreign trade. But the countryside around was still very much as it had been described by Defoe in the reign of Queen Anne, and Pehr Kalm, a young Swedish botanist visiting England on his way to America in 1748, wrote home about the meadows, pastures and market gardens around London and along the river. The gardens were separated by earth walls about four to six feet high, reinforced with rows of oxhorn quicks and covered with grass. The soil was rather thin but heavily manured from the streets of London and from the fields hired to butchers and brewers to turn out their beasts. 'Chelsea is almost entirely devoted to nursery and vegetable gardens', says Kalm. The farmers living near London

> buy and carry home from there all their manure which is collected in the streets and afterwards carried outside the town and laid in great heaps. No farmer who sends in a load of hay to be sold allows his cart to go back empty from the town but it is filled with this manure only, which after it has lain its proper time and fermented into one mass, is spread out in the autumn over the grass in the meadows.[16]

The market gardens of Chelsea and along the river banks were totally inadequate to satisfy the gargantuan demands of the half-million inhabitants of London. Provisions poured in from all over the country, even from as far north as Skye, where the cattle began their long journey south

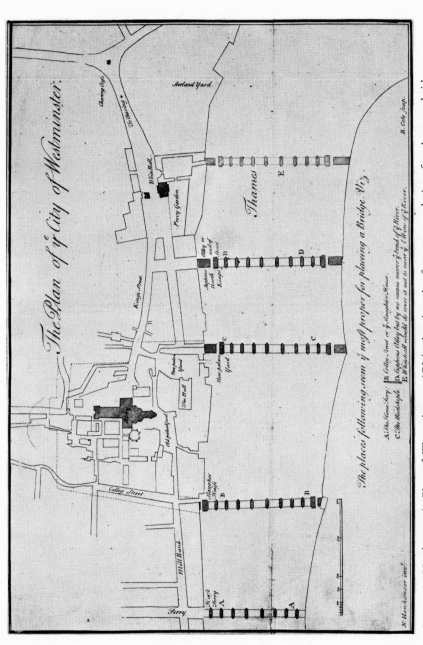

1. Hawksmoor's Plan of Westminster, 1734, showing the five proposed sites for the new bridge. (*Engraving by B. Cole. Photograph by courtesy of the British Museum*)

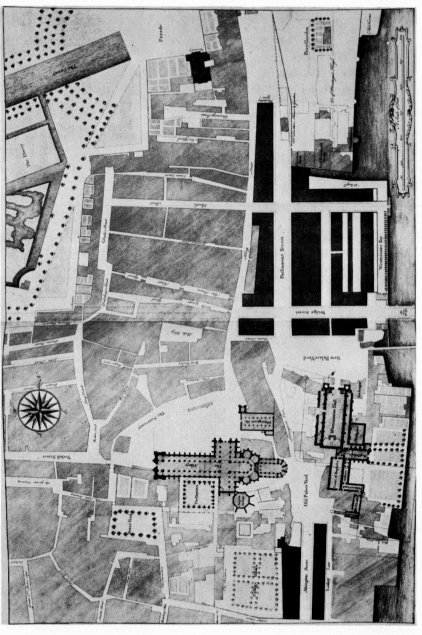

2. Thomas Lediard's Plan of Westminster, 1740, showing the proposed improvements. (Engraving by Fourdrinier. Photograph by courtesy of the British Museum)

by swimming the Kyle of Lochalsh. The main lines of communication along which the supplies travelled were the sea, the rivers, and by pack-horse along the very primitive roads. The Port of London received cargoes from every point of the compass and the myriad vessels thronging into the calm waters below London Bridge had for centuries been a vital part of the life of the country. Defoe, describing the scene in his *Tour*, counted about two thousand sail of all sorts, 'not reckoning barges, lighters, pleasure boats and yachts, but vessels that really go to sea!' But it was down the Thames and the turnpike roads that the greatest volume of traffic descended to supply the town. Before the Duke of Bridgewater and Brindley covered the country with a network of canals during the second half of the century, the rivers of England had been essential as waterways, and efforts to keep them navigable were a constant source of litigation.[17] The importance of the Thames to the needs of London had long been realised and since 1489 authority over its naviga-tion had been vested in the Lord Mayor of London.[18] Various Acts and local orders were enforced to keep it clear of mills, weirs and other ob-stacles, and by 1605 it was navigable almost as far up as Oxford—'the river of Thames is from the City of London till within a few miles of the City of Oxford very navigable and passable with for boats and barges of great content and carriage'.[19] At the end of James I's reign an attempt was made to clear a passage for barges between Oxford and Burcot to enable freestone from the Bullington quarries to be shipped to London—and also so that coal could be conveyed to the University, 'whereof there is now a very great Scarcity and Want'.[20]

The attempt was successful in this respect but towards the end of the century more power had to be given to the Justices of the counties of Wiltshire, Gloucestershire, Oxfordshire, Berkshire and Buckinghamshire, to control the behaviour of barge and boat owners and to fix the exorbi-tant charges of the proprietors of 'Locks, Wears, Bucks, Winches, Turnpikes, Dams, Floodgates and other Engines'. By this time the river was clear as far up as Lechlade, the new Act's preamble stating that

the Rivers of *Thames* and *Isis* have Time out of Mind been navigable from the City of *London* to the Village of *Bercott* in the County of *Oxford,* and for divers Years past from the village of *Bercott* Westward somewhat farther than *Letchlade* in the County of *Gloucester*.[21]

The new control was effective during the Act's existence but the old abuses crept back as soon as it expired until by 1730 conditions were so bad again that petitions had to be presented to the House of Commons by the farmers of Berkshire and Oxford. They claimed that the evils set out in the preamble of the 1695 Act had returned and that 'they now paid 15s for carriage for a Chaldron of Coals from *London* to *Abingdon* for which, within a few Years they paid but 10s, and so in proportion for all other Goods'. Another petition was presented by the Chancellor, Masters and Scholars of the University of Oxford, complaining of extravagant rates imposed by the owners of locks and weirs. Counter petitions were presented by the lockowners protesting that their estates would be prejudiced should the Act be passed.[22] Their protests, typical of many from other rivers, were over-ruled and the new Act was passed.[23]

Further down the river towards Chelsea, Westminster and London Bridge, the water traffic became thicker and more colourful with innumerable passenger boats and sailing craft mingling with the barges and wherries. César de Saussure, writing home in 1725, has left a vivid picture of the Thames at this time.

> It takes its rise in the county of Oxford and is early navigable. The Thames is everywhere wide, beautiful and peaceful; the nearer it flows to London the wider it becomes on account of small tributaries and of the tide, which is felt as far as twelve miles above. You can judge of the width of the Thames by the length of London Bridge which is not built over the widest part of the river yet is eight hundred feet long. You cannot see anything more delightful than this river. Above the Bridge it is covered with craft of every sort; round about London there are at least 15,000 boats for the transport of persons, and numbers of others for that of merchandise. Besides these boats there are others called barges or galleys, painted carved and gilt. Nothing is more attractive than the Thames on a fine summer evening; the conversations you hear are most entertaining, for I must tell you that it is the custom for anyone on the water to call out whatever he pleases to other occupants of boats, even were it the King himself, and no one has the right to be shocked. Some of my friends have told me that on the river Queen Anne was often called 'Boutique d'Eau-de-Vie' because of her well-known liking for the bottle and spirituous liquors. Most bargemen are very skilful in this mode of warfare using singular and quite extraordinary terms, generally very coarse and dirty and I cannot possibly explain them to you.[24]

The bargees have long been notorious for the virility of their language but Dr Johnson is said to have quelled one of them with a counter-attack: 'Sir, under pretence of keeping a bawdy house, your wife is a receiver of stolen goods!' According to a contemporary guide book the coachmen, carmen and watermen in general had the reputation of being 'rude, exacting and quarrelsome' and for the benefit of the public a tariff was laid down. Between all the stairs from London Bridge to Westminster the charge during most of the first half of the century was sixpence by oars and threepence by sculls; from London Bridge to Lambeth and Vauxhall, one shilling by oars and sixpence by sculls; from London Bridge to Hampton Court was six shillings, to Staines ten shillings, and to Windsor fourteen shillings; and the penalty for boatmen demanding more was forty shillings or six months in prison.[25] The boats were numbered and complaints could be lodged at special offices. Any number of these river craft can be seen in the paintings of Samuel Scott and Canaletto, but once again perhaps the most vivid description comes from de Saussure, who delighted in the minutiae of London life.

> Besides hackney coaches and Sedan chairs, London possesses another means of public conveyance in its boats . . . These boats are very attractive and cleanly kept, and are light in weight, painted generally in red or green, and can hold six persons comfortably. On rainy days they are covered with coarse strong tents and in the summer when the sun is burning hot, with an awning made of thin green or red woollen stuff. Some of these boats have two men, called *oars,* others have only one, called *skullers*. You can hire boats in twenty or thirty different places, called *Stairs*. At these places from fifteen to twenty of these Tritons are usually to be found, dressed in a singular sort of doublet pleated about the lower edge, some clad in red, others in green, and on the fronts and backs of their doublets are plates of silver on which are embossed the arms of their masters, for some of these boatmen belong to the King, others to the Prince of Wales and to different peers of the realm, others again to the Lord Mayor or to the magistrates of London. These places are much sought after, for the oarsmen cannot of course join the Fleet in time of war. The boatmen wear a peculiar kind of cap made of velvet or black plush, and sometimes of cloth the same colour as their waistcoats. As soon as a person approaches the stairs they run to meet him calling out lustily, 'Oars, Oars!' or 'Skuller, Skuller!' They continue this melodious music until the person points with his finger to the man he has chosen, and they at once unite in abusive language at the offending boatman.[26]

Rowlandson of course excelled at drawings of quarrelsome boatmen and their customers, and watermen calling out lustily, 'Oars, Oars! Skuller, Skuller!' in the entrance court of Whitehall Palace can be seen in 'Scotland Yard with Part of the Banqueting House', a print from a watercolour drawing by Paul Sandby.[27] These boatmen strenuously resisted the various projects for building bridges over the Thames, petitioning the Commons usually through the Guild of Watermen and Lightermen, and maintaining that any bridge other than London Bridge would mean 'the utter ruin of great Numbers of Families whose sole Livelihood depends on the said Business of Watermen and Lightermen'. [28]

One of the most remarkable features of the eighteenth century in England was the astonishingly swift improvement in the conditions of travel by road. The growth of London and other towns in the British Isles, the increase in foreign trade, changing ways of agriculture and industry, the Union Act of 1707, and many other factors all led to a mobility which amounted almost to a revolution in itself. The replacement of pack-horse trains by waggons and of travel on horseback by carriage and the mail coach meant, sooner or later, the arrival of a sophisticated road system. [29] As London crept further and further westward the business of keeping pace with its insatiable demand for food and drink was a primary cause for the improvement of the road system, including of course the building of bridges all over the country. Defoe's *Tour* is full of accounts of supplies being brought to London in the shape of livestock for the markets. Huge herds of cattle, sheep and pigs, and droves of poultry, cropped their way through the countryside to gather outside the metropolis and await slaughter. Flocks of geese were driven from Norfolk to London where ''tis very frequent now to meet a thousand or two thousand in a drove. They begin to drive them generally in August, when the harvest is almost over, that the geese may feed on the stubble as they go. Thus they hold on to the end of October when the roads begin to be too stiff and deep for their broad feet and short legs to march in'. [30] It can be readily understood that, with all these marching feet and hooves churning up the mud in ever-increasing quantity, the state of the roads, especially in winter or wet weather, could become insufferable. Sussex for some reason, perhaps because of the heavy clay soil, was one of the worst counties, traffic from the south and west sometimes having to travel as far round as Canterbury in order to get to London.

I have narrowly observed all the considerable ways to that impassable county of Sussex, which especially in some parts of the Wild, as they very properly call it, hardly admits the country people to travel to market in winter, and makes corn dear at market because it cannot be brought, and cheap at the farmer's house because he cannot carry it to market . . . I have seen in that horrible county the road, 60 to 100 yards broad, lie from side to side all poached with cattle, the land of no manner of benefit, and yet no going with a horse but at every step up to the shoulders, full of sloughs and holes and covered with standing water. [31]

This was Defoe's account in 1697. Twenty-five years later he was able to write, 'The great Sussex road, which was formerly insufferable bad, is now become admirably good; and this is done at so great an expense that they told me at Strettham that one mile between the two next bridges south of that town cost a thousand pounds repairing including one of the bridges...'[32]

It was easy enough for travellers to exaggerate the state of the roads in order to dramatise their tales with encounters with highwaymen, puddles big enough to drown their horses, boulders as large as houses, and so on; but in general there was no need to exaggerate, for as the traffic increased, and in spite of legislation to try to control the damage done by narrow wheels, the roads remained excruciatingly bad until well into the second half of the century. Petitions poured into Parliament from all over the country. From 1663, when the first Turnpike Act was passed, literally thousands of petitions and separate Acts of Parliament were passed, becoming so tortuous and complicated that by 1766 the situation was intolerable and a comprehensive Turnpike Act was passed. At first there was a certain amount of local opposition, farmers and cattle drovers especially resenting having to pay extra taxes for roads that were not really necessary. The existing roads were good enough and indeed better for cattle than new roads because the stones were liable to cripple them on the way to market. Furthermore, said one petitioner, 'there were 16 Fishermen living near him which go with 8 horses each and that 320 Fishermen's horses go through this road every day ... and between 3 and 4000 Cattle pass thro this Road to London from Sussex which must pay if the Bill passes which will be very burthensome to the Graziers'.[33] Later on, between 1735 and 1750, local protests broke out into rioting, especially in the West Country, where the rioters armed themselves with 'rusty swords, pitch-forks, axes, guns, pistols and clubs', the military had to be called out to keep order, and several Acts of Parliament passed, by which anyone maliciously destroying turnpike gates, locks, flood-gates or other public works, should be judged guilty of felony and suffer death.[34]

A natural accompaniment of road improvement was the building of bridges. Medieval and sixteenth-century bridges were objects of scarcity and even veneration, many of them being maintained by monasteries or a semi-religious fraternity known as *fratres pontifices,* or even by the king himself.[35] After the Dissolution of the Monasteries, the Statute of Bridges was passed in 1531 in an attempt to regulate the responsibilities for their upkeep, in general the liability being handed over to the counties, and in the case of smaller bridges to the parishes.[36] Allowing a bridge to fall into disrepair had always been looked on as a public nuisance penalised by heavy fines, and the Act of 1531 ensured that decayed bridges should not 'lie long without any amendment, to the great annoyance of the King's subjects'. Existing bridges were therefore kept in a reasonable state of repair but it was not until the beginning of the eighteenth century that new bridges began to be built in any number; though probably many of the pack-horse bridges were of seventeenth-century origin, made to cope with the increasing volume of trade and mobility.[37]

Certainly the turnpike roads were responsible for many new bridges, especially over the smaller streams which could flood rapidly after rainstorms, wash away the surface and make sloughs sometimes able to bury both man and horse. 'To give an eminent instance of this', says Defoe in 1725,

> we refer the Curious to take the road from Blackman Street to Southwark in Croyden for an example, where if we are not mistaken, he will find eleven Bridges wholly new built in ten miles length, by which the whole road is laid dry, sound and hard, which was before a most uncomfortable road to travel.[38]

The main activity seems to have been around London and in the Home Counties in order to cope with the quantities of food and supplies coming into the capital. Defoe mentions three hundred new bridges within a few miles of London, 'very plain to be seen'. The method of building these small bridges was observed by Pehr Kalm, ever avid for technical information of this sort.

> I saw nearly everywhere around London that where any water ran under the highway or any other road, they had made a bridge of brick. They dug there a very deep ditch, walled and arched a bridge with bricks and afterwards filled up with earth the part over the arch so that the road was flat and even all the way across it, so much so that there was little sign of any bridge ... There can hardly be any land where larger carts and waggons are used and heavier loads laid on them than in England, where three times as much, if not more, is loaded on a public coach as in Sweden. However I have seen such bridges everywhere in use where I have travelled in England, in Hertfordshire, Bedfordshire, Buckingham, Essex, Middlesex, Surrey and Kent.[39]

The growth of London after the Restoration, the immigration of foreigners into England after the Revocation of the Edict of Nantes, the increase of building following the Peace of Utrecht and the end of the Marlborough Wars, the agrarian revolution, the growth of trade both at home and abroad, all these were factors making an improved system of communications a vital necessity to the life of the country. The roads were getting better, the rivers were being cleared and made navigable, a canal system was very shortly to come. But for centuries the Thames below Staines and Kingston had no bridge until the river flowed under the ancient and dangerous structure of London Bridge.

## NOTES TO CHAPTER 1

1. Acts 13 & 14 Charles II, c 2 (1662).
2. London Topographical Society 15 (1904).
3. Evelyn *Diary* 15 October 1664; Pepys *Diary* 20 February 1665.
4. M. D. George *London Life in the Eighteenth Century* (1925) ch 2; George Rudé *Hanoverian London 1714–1808* (1971) ch 1.
5. Grey *Debates* 23 January 1673/4.
6. CJ 4 March 1709/10, vol 16 p 345.
7. Bridge Minutes 7 November 1739.
8. César de Saussure *A Foreign View of England in the Reigns of George I and George II* (1902 edition) p 67.
9. Stow described King Street as 'somewhat narrow which causeth stoppages by reason of Multitudes of Coaches' (*Survey* IV p 63).
10. Kennel (canal, channel), a gutter usually in the middle of the street. Hogarth's 'Noon' shows a dead cat floating down the kennel in Hog Lane, Westminster.
11. CJ 19 February 1728/9, vol 21 pp 229–30.
12. Act 2 George II, c 11 (1729); see also 2 George III, c 21 (1761), 3 George III, c 23 (1762), and 4 George III, c 39 (1763).
13. Bridge Minutes 21 November 1739.
14. Vestry Minutes of St Margaret's Westminster, 12 February 1739/40 pp 41–3.
15. Lord Tyrconnel in a debate in the Commons 27 January 1740/1 (*Parl. History* vol 11 pp 1010-13) taken from Dr Johnson's account in the *Gentleman's Magazine* XII pp 179–81.
16. Pehr Kalm *Visit to England* 1 and 10 May and 24 June 1748.
17. W. T. Jackman *The Development of Transportation in Modern England* (1962) ch III.
18. Act 4 Henry VII, c 15 (1489), and the first Charter of James I, 20 August 1605.
19. Act 3 James I, c 20 (1605/6).
20. Act 21 James I, c 32 (1623); F. S. Thacker *The Thames Highway* (1968) ch IV, 'The Oxford-Burcot Commission'.
21. Act 6 & 7 William III, c 16 (1695).
22. CJ 26 January and 9 March 1729/30, vol 21 pp 417 and 485.
23. Act 3 George II, c 11 (1730).
24. De Saussure op cit pp 93–5, 16 December 1725.
25. Chamberlayne *The Present State of Great Britain* (1741) p 262.

26. De Saussure op cit pp 169–70.
27. Published by Edward Rooker after Sandby 31 December 1766, see London Topographical Society Record VI (1909) pp 37–8.
28. CJ 23 January 1721/2, vol 19 p 724.
29. Sidney and Beatrice Webb *The Story of the King's Highway* (1913); W. T. Jackman op cit ch II; H. L. Beale 'Travel and Communications' in Turberville *Johnson's England* (1933) ch VI; George Rudé *Hanoverian London 1714–1808* (1971) ch 1.
30. Daniel Defoe *A Tour through the whole Island of Great Britain* (1724–7) I pp 88–9.
31. Daniel Defoe *An Essay on Projects* 1697 (Carisbrooke Library edition p 59).
32. Defoe *Tour* II p 191.
33. CJ 10 February 1724/5, vol 20 p 405.
34. Acts 1 George II, c 19 (1728), 5 George II, c 33 (1732), 8 George II, c 20 (1735) and 27 George II, c 16 (1754), the last making the former Acts perpetual.
35. For the difficult problem of legal responsibility for the maintenance of bridges, see W. S. Holdsworth *History of English Law* X (1938) pp 322–32.
36. Act 22 Henry VIII, c 5 (1531). For an abstract of Statutes on County bridges see *Report of the Middlesex Magistrates on Public Bridges* (1826) pp 22–35.
37. Webb op cit ch VI, 'The Maintenance of Bridges'.
38. Defoe *Tour* II p 192.
39. Pehr Kalm *Visit to England* 26 April 1748.

# 2
# Early Attempts

Until 1729, when Fulham Bridge was built, apart from ferries the nearest crossing up-river from London Bridge was the ancient timber bridge at Kingston. London Bridge needs no description here.[1] With the ever-increasing traffic coming in and out of London the jams on the bridge were becoming insupportable and in 1722 an Order was published by the Common Council instructing traffic to keep to the left.

> All Carts, Coaches and other Carriages coming out of Southwark into this City do keep all along on the West Side of the said Bridge . . . and that no Carman be suffered to stand across the Bridge or load or unload . . . and that Duties be collected without making a Stay of Carts.[2]

The drawbridge was repaired in the same year but the old gateway was allowed to stand, forming a bottleneck eighteen feet wide.

Shortly before his death in 1736 Nicholas Hawksmoor, Clerk of the Works at Whitehall and the senior architect in the profession, produced two schemes for London—one a major alteration to London Bridge, the other a new bridge at Westminster. London Bridge, said Hawksmoor, besides being inconvenient and dangerous, as an architectural construction was positively barbarous and had changed 'the gentle Current of the Thames into many frightful Cataracts, which very much obstructs and indangers the Navigation thro' the Bridge, as is often experienced by the Loss of Lives and Goods, and the Vessels which are either thrown upon the Sterlings or sunk within the Arches'. He proposed to demolish the four central arches, leaving the middle pier as an anchor point and building from it two large arches about the same breadth as the great arch at York. This would allow the tide to flow freely and he believed that shipping masters would willingly pay tonnage for 'such a Conveniency and Security of their Goods and Vessels'.[3] He addressed his suggestion to the Members of Parliament for Westminster, Lord Sundon and Admiral Wager, but he anticipated opposition from the City and indeed nothing further was heard of the project until 1746 when a number of meetings were held at the Guildhall to try to decide the bridge's future.

Charles Labelye, the Swiss engineer to become famous as the architect of Westminster Bridge, proposed two schemes, both aimed at reducing what he considered the main defect of London Bridge, 'the monstrous largeness of the present Sterlings'. His first plan was simply to replace the sterlings with Portland stone piers and thus to double the waterway to about 400 feet. The second plan he admitted to be the idea of Sir Christopher Wren who, as part of the post-fire Thames Quay scheme, had intended to remove every alternate arch and to reduce the piers from nineteen to ten, affording a waterway of about 540 feet. Labelye's new arches were to be Gothic and to spring from the lowest Low Water Mark, thus adding greatly to their strength. The superstructure was to consist either of new houses or of a cast-iron balustrade on a dwarf parapet with Gothic recesses over each pier. 'The expense', concluded Labelye, 'would not be much more than the other, and *London Bridge* so mended would be the finest and most commodious *Gothick* Bridge in the World.'[4] A number of other witnesses were examined including the Clerk of the City Works, George Dance, the tide carpenter, Bartholomew Sparruck, two master masons, Christopher Horsenaile and Andrews Jelfe, and Francis Hawksbee FRS, but the deliberations came to nothing and it was not until the Act of 1756 empowered the City to demolish the houses on the bridge that the attempts to relieve traffic congestion were anything like successful.[5] By 1762, as a result of the Act and the labours of a committee appointed to advise on the possibilities of widening the bridge, the famous houses, 'a public nuisance, long felt and universally censured and complained of', were pulled down and the width of the bridge was increased to forty-six feet, with lamps and domed pedestrian shelters. The work was carried out by George Dance and Sir Robert Taylor adopting a feeble dilution of Hawksmoor's idea. They replaced the two central arches with a large single arch which had the effect of allowing a tremendous volume of water to pour through the middle, scour the bottom in the pool below, and irretrievably weaken the foundations.[6] In spite of quantities of stone from the City walls packed round the sterlings and into the erosion below the middle arch, the trouble was never satisfactorily cleared up until Rennie's new London Bridge was begun in 1824.

For centuries the first bridge across the river for twelve miles above London was at Kingston-on-Thames. The history of Kingston Bridge is as long as, if less melodramatic than, that of London Bridge.[7] A timber bridge is believed to have been built there after the Roman invasion and to have been slightly lower downstream, but the bridge which existed there in the eighteenth century was probably of Saxon origin and built further up for the convenience of the new Saxon town of Moresford. It was endowed with land for its maintenance in 1219 but was in a state of dilapidation in the reigns of Henry III, Edward III and Henry VI, the repairs being carried out by bridge-wardens and by the bailiffs and freemen of Kingston. Henry VIII ordered the 'repair of the Great Bridge' in order to save London Bridge from damage by his heavy artillery, and Mary granted forty perches of water in two fish weirs for its upkeep. Finally in 1565 Robert Hamond, bailiff of Kingston, bequeathed £40 annually from his estate at Harmondsworth 'to the intent that every person might pass over the said Bridge freely', and this was ratified by his son Robert who recorded the fact on a brass plate in the middle.

John Aubrey described Kingston Bridge at the end of the seventeenth century when, though many times damaged and repaired, it was still basically the same as the Saxon original. 'The great wooden Bridge over the Thames', says Aubrey in the *Perambulation of Surrey*,

> hath twenty-two Piers of wood that support it, two great interstices for Barges to pass through, which twenty-two Piers do contain 126 Yards; besides at the East End thirty yards of Stone and Brick, and at the West End twelve yards, which contain in all 168 Yards. In the middle of the Bridge are two fair Seats for Passengers to avoid Carts and to sit and enjoy the delightful Prospect.[8]

Attached to the east end was a ducking stool, a basket on the end of a beam used for punishing 'those turbulent women who did not properly understand the regulation of their tongues'. An account of its use was published in the *London Evening Post* on 17 April 1745: 'Last week a woman that keeps the Queen's Head Alehouse, Kingston, was ordered by the court to be ducked for scolding and was accordingly placed in the chair and ducked in the river Thames under Kingston Bridge in the presence of two or three thousand people'.[9]

Above Kingston there was a ferry from Chertsey to Littleton until the fourteenth-century bridge was built by Chertsey Abbey and noted by Leland as 'a goodly bridge of timber newly repaired'.[10] In the eighteenth century, at the time of the projected Fulham Bridge, Captain Perry surveyed the bridges at both Kingston and Chertsey,

> which require (as all other Bridges built in like manner) frequent, expensive and tedious Repairs, and all Passage by wheels and otherwise whilst such

repairs are carrying on, are under the necessity to be perform'd by the incommodious and tedious use of Ponts or Ferry Boats.[11]

Chertsey Bridge was rebuilt in stone by James Payne and Kenton Couse in 1780-5 at the expense of both Surrey and Middlesex. Kingston Bridge was rebuilt in brick with stone facing in 1828.

The main crossing, however, was at Staines where, as its name *Pontes* implies, there had been a bridge from Roman times and probably even earlier.[12] By 1740 Staines Bridge was in 'a ruinous and dangerous condition and the Causeway fallen to such decay that unless speedily repaired it will be in great danger of being broken through by high Floods . . . and the road from Staines to Egham, being part of the Great Western Road, rendered impassable'.[13] By 1791 it was thought to be 'narrow and incommodious and so greatly decayed that in the opinion of able and experienced workmen, it ought to be taken down and a new Bridge built'.[14] Another bridge was designed immediately by Thomas Sandby but it was defective and had to be replaced shortly after its architect's death in 1798.[15]

Between Staines and Wallingford the bridges were frequent and evenly spaced but mostly built of timber.

The normal method of crossing the river was by ferry; many existed from Tilbury upwards, capable of carrying horses, cattle and coaches. They were erratic, uncomfortable and often exorbitant in their charges. Pepys describes a crossing in the grim Plague year of 1665:

Up and very betimes by six o'clock at Deptford and there find Sir George Carteret and my Lady ready to go, I being in my new coloured silk suit and coat trimmed with gold buttons and gold broad lace round my hands, very rich and fine. By water to the Ferry, where when we come, no coach there, and tide of ebb so far spent as the horse-boat could not get off on the other side the river to bring away the coach. So we were fain to stay there in the unlucky Isle of Dogs in a chill place, the morning cool and wind fresh, above two if not three hours to our great discontent. . . Sir George the most passionate man in the world did bear with it and very pleasnt all the while.[16]

Not many had regular time-tables and some were forbidden to ply between sunset and sunrise without special permission.[17] In 1699 there were seventeen plying places between Vauxhall and Limehouse, two of them being at Westminster. One of these was the Lambeth Horseferry, belonging to the Archbishop of Canterbury. This had a special arrangement for Sunday ferrying by which the waterman took it in turns to 'ply and work on Sundays across the River from and to *Westminster-Bridge* and

Stangate and the *Horse-ferry* at *Lambeth,* which Money so earned is applied to the Use of the poor Watermen or their Widows of the Parish of St Margaret'.[18] The Lambeth Horseferry however remained notorious from every point of view and one of the reasons for building Westminster Bridge was to overcome this trouble. As Hawksmoor said in his prospectus for the new bridge, 'there is no need of saying anything of the Badness and Inconveniency of *Lambeth-Ferry,* since there is scarce anyone ignorant of it, and some have found it to their Cost'.[19]

A constant source of irritation to ferry passengers was the behaviour of the ferrymen whose bad manners and filthy language were a byword. The Company of Watermen and Lightermen went to considerable trouble to control this nuisance. An order of 1701, for instance, noticed that the watermen 'do often use such immodest, obscene and lewd Expressions towards Passengers and to each other, as are offensive to all sober Persons, and tend extremely to the corrupting and Debauchery of Youth', and imposed a fine of 2s 6d for each offence.[20] The swearing continued, however, despite the penalties, but Rowlandson's drawings show that the passengers were probably almost as skilled in abuse as the watermen themselves.

Another attempt made by the authorities to protect the public tried to insist that men in charge of boats should be reasonably responsible. One of the many Acts passed in the eighteenth century for bettering, ordering and governing the watermen ordered that 'no Apprentice shall take upon himself the Sole Care and Management of any Boat . . . till sixteen Years of Age if a Waterman's Son, and seventeen if a Landsman's, and unless he has worked with some able Waterman for two Years at least, on Pain of 10s on the Master'.[21] This measure was more successful; except for the hazardous and foolhardy sport of shooting the rapids of London Bridge, and for occasional disasters caused by the submerged works of Westminster Bridge, serious accidents were relatively few.

Confusion is sometimes caused by the use of the word *bridge* in various documents and maps though the meaning is clearly that of a landing-stage for boats.[22] Lambeth Bridge, sometimes known as the 'Great Bridge', was a magnificent timber landing-stage capable of receiving the monarch on state occasions, notably the meetings between Henry VIII and Cranmer, and Queen Elizabeth and Archbishop Parker. Henry Chichele's account books in Lambeth Palace contain bills for the repair of Lambeth Bridge which is described as being strongly built of wood with a chain and two iron posts at the inner end and a flight of steps to the water at the other end.[23] The word has certainly been used in this connection since the fourteenth century, when Edward III granted duties on wool and leather in order to build a *bridge* for the new Woolstaple at Westminster. Stow speaks of 'New Palace Yard ... here is *Westminster Bridge* for taking boat, for such as are minded to go to London or elsewhere by water';[24] and the Middlesex magistrates ordered John Lee and William Bishop to be set in the stocks 'for emptying a privie and casting the excrement at the King's Bridge at Westminster: and a paper be set upon their heads expressing their offence'.[25] A small corporation of twenty-four men, the Society of Bridge Porters of Westminster, was licensed by the Burgess Court to 'carry all Burthens of Corn, Mault &c to and from the said Bridge', their privilege being jealously guarded against encroachment, especially from the City of London porters.[26] By 1733 the King's Bridge at Westminster was in such a state of dilapidation that it was unsafe and the Treasury ordered the Board of Works to repair it at the cost of £250.[27]

These bridges and many others were all quays or jetties built out on piles a short way into the river, usually with a floating stage to allow for the tide. They developed into elegant stone constructions with flights of stairs leading down into the water, sometimes with gateways and balustrades. The bridge to the river front of Somerset House, probably by Inigo Jones, can be seen in Canaletto's painting at Windsor Castle (Plate 4).

A pamphlet, published anonymously in 1738 and describing the origins of Westminster Bridge, mentions various attempts in the reigns of Elizabeth I, James I and Charles I to project a bridge at Westminster.[28] According to Sir Henry Herbert, Master of the Revels to both James I and Charles I, there had been several endeavours before the Civil Wars to build bridges at Lambeth and Fulham, and after the Restoration the idea was put up again but as usual monopoly interest proved too strong.[29] On 17 August 1664 in Whitehall Palace the king presided over the Privy Council attended by the Lord Mayor of London, the bailiff and magistrates of Westminster, the master and wardens of the Company of Watermen and the farmers of Lambeth Ferry, to discuss

> whether it would not prove of good Advantage that a Bridge should be built over the River of Thames between Westminster & Lambeth for the better communication of Southwark & those parts with the Cities of London & Westminster.

It was ordered that they should meet again on the Wednesday before Michaelmas to bring their reasons for and against the proposal.[30] As this happened to be in fact St Michael's day, when the sheriffs for the following year were due to be sworn in, the meeting was postponed until 5 October.

Meanwhile the City lost no time in drawing up a draft setting out their reasons why a bridge should not be built. London Bridge, they said, was perfectly adequate and another would be positively harmful to the whole country, endangering the king and his family in case of insurrection, diverting trade from Southwark and plunging 8,000 families living there into poverty and discontent. The inhabitants of London Bridge and the south part of the City would suffer too, and indeed London Bridge itself would certainly fall into decay because daily repair and great expense was necessary for its support. Furthermore a new bridge would ruin trade between London and the West Country because no one would dare to brave the violent current it would cause except possibly at slack water. Nor would barge owners allow their craft to pass beneath in case the piles be swept away and the bridge collapse. The river would rise at least three feet higher and much more at spring tides, drowning the lower rooms of Whitehall Palace and flooding the cellars, warehouses and dwellings on both sides of the river as far up as Chelsea and Battersea. In fact the river itself would silt up and become no longer navigable to the great evil of the 30,000 watermen earning their living on it, 'besides considering how great a Nursery of Seamen this hath been and how ready and useful they have appeared in the late Warres with the United Provinces'. Finally the City declared that a new bridge would increase the morass of building along the river which would be further polluted by sewage, soil and filth, and they implored the king to 'give all just discouragement to the proposal leading so apparently to the decay and danger of the most ancient City of London and by consequence of the whole Kingdom'. [31]

This specious and myopic plea was typical of many others to follow and it was countered with a battery of arguments advanced by a spokesman for the Westminster citizens and Home County landowners. He airily demolished the City's objections and then went on to list his own reasons in favour of a bridge. The first, with which he no doubt hoped to enlist the sympathy of Charles II himself, was that a new bridge at Westminster would open up a land route between the various palaces at Hampton Court, Whitehall, Nonsuch and Greenwich, not to mention the Duke of York's house at Richmond. Greenwich was specially relevant at that time because of the fine new house being built there by John Webb for the queen. He mentioned that a point worth considering was that horses which arrived steaming hot at the ferries often had to wait two or three hours for a passage. His second reason was the increase of trade a bridge would bring to Westminster 'which is very poor and very populous and depends much upon the Court'. The third was a military and political precaution.

> The Borough of Southwark [he warned], being the nest of the Fanaticks insomuch as within these 2 or 3 days they assembled no less than 4 or 5000 and if they should grow to that insolence that it were necessary to send the Guards to suppress them, it is at least three hours march thither . . . where if there were a Bridge wee might quickly bee upon them and their apprehension of being attacked on all sides will much contribute to the keeping them in better order.

He also proposed that a strong quarter to lodge the troops should actually be built on the bridge itself.[32] Southwark, a notoriously disorderly district with its theatres, inns, bear gardens and bowling-alleys, had long been a haunt of discontent and conspiracy and a plan to cut an easy way into it might well have been a sensible move appealing to Charles II's sense of tactical expediency in safeguarding his kingdom.

However all these reasons, patently for the public good though they may have seemed, were over-ruled by an unanswerable argument typical of Charles's hand-to-mouth fiscal expedients. It took the shape of a loan of £100,000 raised by the City, according to Pepys, 'without any security but the King's word which was very noble'.[33] This bribe proved effective enough to silence the claims either for or against the bridge and the City was moved to record its gratitude to the king for his support:

> it is left to the Right Hon. the Lord Mayor and Alderman to acquaint His Majesty with the City's ready compliance and dutiful submission towards His Majesty's pleasure and withal to present unto His Majesty the City's great sense and Apprehension of and most humble thanks for the great Instance of His Majesty's goodness and favour towards them as expressed in preventing of the new Bridge proposed to be built over the River of Thames betwixt Lambeth and Westminster which as is considered would have been of dangerous consequence to the State of this City.[34]

Pepys knew of the failure of the project in October but curiously enough, two months earlier, the Dutch Ambassador, Van Gogh, had written home to the States General that 'a project brought before Council for a new bridge between Westminster and Lambeth is laid aside from opposition of the Londoners', and a similar report had been sent by the Venetian Ambassador, both of them no doubt shrewdly judging the atmosphere at

3. Detail from Rocque's *Survey of London*, 1746, showing Westminster and Lambeth marshes. (*Engraving in the GLC Record Office. Photograph by R. B. Fleming*)

4. Somerset House Stairs (or Bridge). (*Detail of a painting by Canaletto in the Royal Collection, reproduced by gracious permission of Her Majesty the Queen. Photograph by A. C. Cooper*)

the first meeting with the king in August.[35] The City magnates, assessing accurately enough Charles's chronic impecuniosity and well aware—mistakenly, as it turned out—of where their own interests lay, did not hesitate to use their immense power to scotch a project which could have been of benefit to the country from every point of view, social, commercial, military and aesthetic. For a seventeenth-century bridge at Westminster, built perhaps by Wren and incorporated in the rebuilding of London after the Fire, might have become an architectural feature worthy to rank with the City churches and St Paul's, and of incalculable value to William III and Marlborough in the wars against Louis XIV.

The formidable opposition mounted against the idea of any rival to London Bridge can be seen reflected in the difficulties met by the promoters of the bridge between Fulham and Putney, a strategic point half way between the City and Kingston-on-Thames, where a ferry had been recorded since Domesday Book and where a military pontoon bridge had been set up by the Earl of Essex against the armies of Charles I. Projects to build bridges at Lambeth and Fulham before the Civil Wars had all failed. The first serious attempt to build bridges at these two points is recorded in 1670 when plans were bitterly opposed by both the Corporation of London and the Watermen's Company.[36] A debate in the House of Commons in 1671 on a 'Bill for building another Bridge over the River Thames from Putney' was led off by the City Member, John Jones, who in an almost hysterical flood of verbiage stated the usual City objections.[37] He believed the Bill would question the very being of London and 'next to the pulling down of *Southwark,* nothing can ruin it more. All the correspondences westward, for fuel and grain and hay, if this bridge be built cannot be kept up'. Jones was supported equally vehemently by the serjeant-at-law, Sir William Thompson, who said, 'This will make the skirts (though not *London*) too big for the whole body; the rents of London Bridge, for the maintenance of it, will be destroyed. This will cause sands and shelves and have an effect on the Bridge Navigation and cause the ships to lie as low as Woolwich; it will affect your navigation, your seamen and your Western barges who cannot pass at low water'. Hugh Boscawen, Member for Cornwall, thought that if a bridge were built at Putney they would soon have others at Lambeth and elsewhere to the ruin of the watermen and against the wishes of the counties of Surrey and Middlesex. Sir Thomas Lee, rather more cynically, suspected the only reason for putting the Bill forward was to increase the value of property westwards. A more rational view was stated by Edmund Waller, Member for Hastings, poet and a parliamentarian of varied loyalties, who spoke up for the bridge emphatically. 'If ill for Southwark', he said, answering the City objections of Jones, 'it is good for this end of the town, where Court and Parliament are. At Paris there are many bridges, in Venice hundreds. We are still obstructing public things. The King cannot hunt but he must cross the water. He and the whole nation will have convenience by it.' Burnet described Waller as 'the delight of the House and even at eighty he said the liveliest things of any among them', but he was over-ridden on this occasion in spite of the support of Sir John Trevor, secretary of state, and Colonel Stroud, who regarded any bridge as an asset from a military point of view. Vested interests prevailed and the Bill was knocked on the head by Sir Henry Herbert's plea that the project was of considerable public interest and the House too thinly represented on that occasion for a Bill to be introduced. It was rejected by sixty-seven votes to fifty-four. [38]

No further action was taken until Sir Robert Walpole is said to have been held up on his way to the House of Commons by the Fulham ferry being high and dry on the shore and the ferrymen drinking, oblivious to his shouts, in the Swan Tavern. A petition was presented to the Commons by the freeholders and inhabitants of Middlesex, Surrey and Sussex, praying for a bridge either from the Cage at Chelsea to Lord St John's house at Battersea (the site of Battersea timber bridge, built in 1771-2), or between Hurlingham and Wandsworth or Fulham and Putney. [39] Counter-petitions were presented by the Corporation of London, the governors of St Bartholomew's Hospital, the justices and landowners of Southwark, the Fulham and Putney ferrymen, and by Thomas Osborne, the tenant of the ferry between Chelsea and Battersea, all claiming that a bridge at Fulham would impair their trade and cause hardship and even ruin to their families. [40] The House examined these petitions fairly closely and most of the witnesses called to give evidence testified to the hazards of the Fulham ferry. Abraham Odell, for instance, said that the passage was very dangerous and he personally had seen 'the Ferry-boats, in windy and tempestuous Weather, drove a great way down the River and the People have been in danger of being lost; and there were three Persons drowned about three Months agone; and that he had seen Coaches and Carts ferry over and the Horses often break their Braces and have been in danger of being drowned; and that sometimes the Coaches can't get over at all'. Thomas Scott said that ferries were often driven a long way downstream and that horses suffered from long waits at the ferry. William Cross, one of the Fulham ferrymen, said the ferries were sometimes driven as far down as Wandsworth; and Robert Grey, landlord of the Swan, said that only a few days ago he had been obliged to get boats to rescue a coach and horses in danger of being lost with their passengers. He said the river at that point was shallow and easily whipped up by the wind, and sometimes he had even been able to wade across. These witnesses probably made the most of their stories but they were supported by the Comptroller of the Works, Thomas Ripley, who said that he had taken borings and measurements in various places and thought that a bridge could easily be built either at Chelsea or at Fulham and 'would in no wise affect or prejudice Navigation . . . and that a Bridge would be a publick good'.[41] The Commons made various amendments and finally passed the Bill on 10 May 1726. A further Act was passed in 1728 as a result of the commissioners petitioning for more powers to deal with building costs and for compensation to the Fulham Horseferry and to the Sunday ferrymen.[42]

Two hundred commissioners, Members of both Houses of Parliament, were appointed to lay out the bridge and its approaches and to decide on how best it could be financed, their first meeting being held on 26 July 1726 at the Swan Tavern.[43] They were incorporated by Royal Charter and after a certain amount of deliberation it was decided to finance the bridge by means of a body of thirty subscribers contributing £1000 each. The list was headed by Sir Robert Walpole himself and included Admiral Wager, Ripley, Price, Sir Matthew Decker, Captain Perry, Dr Cheselden and Lord Carpenter, MP for Westminster.[44]

At the next meeting, on 22 August, the first designs were submitted by Thomas Ripley, John Price and William Halfpenny. Ripley's design was for an oak and yellow fir timber bridge on oak piles, twenty-three feet wide, to cost £8,000, to be finished in two months and to last for at least twelve years. A bridge built completely of oak should last for thirty years and could also be used as scaffolding for a stone bridge. Plan two was submitted by John Price, an architect previously employed by the Duke of Chandos at Cannons House, and was to be twenty feet wide, to cost £8,600, to be finished in nine months and to last for fifty years. He did not approve of Ripley's design and did not believe a timber bridge could be used in any way as scaffolding for a stone bridge. Plan three was by William Halfpenny, sometimes known as Michael Hoare, a master carpenter of Richmond and author of a number of architectural pattern books. His design was more ambitious, costing £28,000, could be built in one year and would last for thirty years. The committee accepted Ripley's design with a few modifications, and also Thomas Phillips's estimate for timber construction and Andrews Jelfe's for Portland stone courses.[45] But for some reason not clear in the Minutes the design was not executed and, apart from settling compensation claims, the scheme lapsed for about two years. Then in July 1728 the commissioners inserted advertisements in the *Daily Courant* for new designs to be submitted to the Secretary at the Lottery Office in Whitehall. At first only two letters arrived, one from Richard Newsham of Cloth Fair, the other from Captain J. Timothy Perry who was then engaged on work in Lincolnshire and in Rye harbour. Perry engaged to build a bridge 'by a method not yet practised in England', no doubt drawing on his experience as a civil engineer in Russia. After a few weeks he produced a model. John Price then reappeared with the model of a stone bridge which he had submitted in 1726, to be built for £45,000; and Ripley turned up again with designs for a stone bridge with timber arches.[46] Then at a meeting at the Whitehall Lottery Office on 27 November 1728, models and drawings were produced by Ripley, Price, William Halfpenny, Goodyear, Captain Perry, and two designs by Sir Jacob Ackworth who also showed the committee drafts of the timber bridges at Kingston, Chertsey, Staines, Datchet and Windsor for comparison. At the next meeting, in December, Admiral Wager reported that after considering a number of models and designs and with the help of Sir William Ogborn and Sir Jacob Ackworth, his committee had selected two designs: a timber bridge by Ackworth (£9,455) and a moor-stone and timber bridge by Ripley (£15,000). Achworth's estimate was handed in at the end of January and the work was given to the master carpenter, Thomas Phillips. The final cost of building old Fulham Bridge amounted to £23,084 14s 1d. The first piles were driven in March 1729, and in October Colonel Armstrong was desired to go to Fulham and view the completed bridge in his professional capacity of Surveyor General. He reported at the next meeting that he had examined the new bridge and had found that 'it is built within the limits prescribed by the Acts of Parliament and upon better principles than the draught signed by the Commissioners . . . and the work is well executed'.[47]

In fact, once the preliminary arrangements had been settled, the actual construction took about eight months. The charming story that it was designed by William Cheselden, one of the subscribers and a celebrated surgeon of St Thomas's Hospital—the right man to construct a bridge, with its many wooden legs—is given in Faulkner's *History of Fulham,* and the Bridge Minutes confirm this. There is no indication when the transfer from Ackworth's design took place but at a meeting on 2 July 1730 a vote of thanks was recorded by the proprietors making Cheselden's authorship fairly clear.

Resolved, as the bridge is built entirely according to a scheme and principles laid down by Mr Cheselden, and as he has been very serviceable in directing the execution of the same, that the thanks of the Proprietors be given him for the advantages which have been received from his advice and assistance, they being of the opinion that no timber bridge can be built in a more substantial and commodious manner than that which is now erected.[48]

The settlement of claims for compensation, compared with the difficulties to follow over the ownership of land at Westminster Bridge, were decided relatively easily. Interests in the ferry were vested in the Manors of Fulham and Wimbledon, held by the Bishop of London and the Duchess of Marlborough. They were settled with a sum of money and the right to cross without toll. The copyhold tenants, Daniel Pettiward, William Skelton and Hamon Gotobed, after a certain amount of protest at the alleged loss of their livelihood, agreed on £8,000 between them. And the watermen and their families received compensation by Act of Parliament amounting to £62 in perpetuity to be paid annually by the churchwardens of Putney and Fulham.

Walpole was immortalised in the centre lock, named after him and his exertions at the beginning. But the bridge itself seems to have had its drawbacks, being built at an angle and curved at the Putney end. The flood tide at that point sets strongly towards the Putney shore and as Dr Desaguliers reported to the Commons in 1735, 'although it is but little oblique to the Stream it has occasioned the Loss of many Barges till the Bargemen learned the skill of overcoming these Difficulties'.[49] During the hard frost in 1757 ice demolished one of the wooden piers, terrifying a coachload of gentlemen,[50] but paintings of the bridge show an extremely solid structure with an eighteen-foot carriage-way, two coaches passing with plenty of room to spare and several spacious pedestrian triangles. It was renamed Putney Bridge in the nineteenth century and demolished in 1881 to be replaced with the present stone bridge by Sir Joseph Bazalgette.[51]

Fulham Bridge marked an important advance in the development of communications. As James Bramston said in *The Art of Politicks,*

> Our Fathers crossed from *Fulham* in a Wherry,
> Their Sons enjoy a Bridge at *Putney-Ferry*.

Apart from it being built of timber and financed by private enterprise instead of a government lottery and Parliamentary grants, there are other striking similarities between Fulham Bridge and the new bridge at Westminster begun ten years later. They both met with fierce opposition from the City and the watermen, and in each case local endeavours to build bridges shortly after the Restoration had been defeated by the selfishness and prejudice of the Corporation of London. They were both promoted to varying extents by Walpole and Admiral Wager and many of the commissioners served on both Bridge Commissions. Compensation had to be paid in the one case to the Bishop of London, in the other to the Archbishop of Canterbury, for their interest in the horseferries. And they both proved their worth instantly as key points in the growing network of roads throughout the south of England. Fulham Bridge was in many ways a testing ground and can be examined as a forerunner to the more ambitious project which was to follow lower down the river.

## NOTES TO CHAPTER 2

1. Among the copious literature on London Bridge, see Richard Thomson *Chronicles of London Bridge* (1827) and Gordon Home *Old London Bridge* (1931).
2. Strype's *Survey* (1734–5) Bk I p 49.
3. Nicholas Hawksmoor *A Short Historical Account of London-Bridge* (1736) p 13.
4. Maitland *History Of London* (1774–5) II pp 837–33 with an engraving by B. Cole.
5. Act 29 George II, c 40 (1756).
6. See an engraving of the timber centring for the arch in BM *King's Maps* xxii 36 e-i.
7. *Report of a Committee of Middlesex Magistrates on Public Bridges* (1826) p 284; W. D. Biden *History of Kingston-on-Thames* (1852) pp 59–63; VCH *Survey* III (1911) pp 488–9.
8. John Aubrey *Perambulation of Surrey* (1673) printed in *The History of Surrey* (1719) I pp 44–6.
9. Henry Humpherus *History of the Company of Watermen and Lightermen* (1859) II p 194.
10. Leland *Itinerary* (1535) I fol 123. The bridge is illustrated in VCH *Surrey* III, plate facing p 404.
11. Captain Perry's letter to the Fulham Bridge Commissioners, 24 September 1728 (Fulham Bridge Minutes p 103).
12. Susan Reynolds, VCH *Middlesex* III pp 13–14.
13. Act 13 George II, c 25 (1740).
14. Act 31 George III, c 84 (1791).
15. Sandby was the designer of a 'Bridge of Magnificence' to cross the Thames from Somerset House, 'solely to emphasise the value of symmetry and to create an impression of grandeur'. It was never intended to be built.
16. Pepys *Diary* 31 July 1665.
17. F. S. Thacker *The Thames Highway: Locks and Weirs* (1920) p 495.
18. Acts 11 & 12 William III, c 21 (1700).
19. Hawksmoor op cit p 16.
20. Order of the Company of Watermen, 8 October 1701, cited by Maitland *History of London* II p 1255.

21. Act George II, c 26 (1729).
22. J. G. N. in *Gentleman's Magazine* (1852) pp 486–7 and (1855) p 554.
23. Lambeth Court Rolls No.549 (2–3 Henry VI) cited by Dorothy Gardiner *The Story of Lambeth Palace* (1930) p 46.
24. Stow's *Survey* (1720 edition) Book VI p 62.
25. Middlesex Session Books 25/25, January 1640/1.
26. Burgess Court Records 9 December 1707 cited by W. H. Manchee *The Westminster City Fathers* (1924) pp 78–82.
27. *Calendar of Treasury Books and Papers* (1731–4) p 375.
28. *A Short Narative of the Proceedings of the Gentlemen concerned in Obtaining the Act for building a Bridge at Westminster* (10 November 1737) *With a Member of Parliament's Answer* (24 November 1737).
29. Sir Henry Herbert in the House of Commons 4 April 1671.
30. Letters between the Privy Council and the City, *Remembrancia* IX 96–7, and rough draft in Guildhall MSS 119.8.
31. Letter from the City to the Privy Council 19 October 1664, *Remembrancia* IX 98 and Guildhall MSS 119.8.
32. 'Concerning the Proposition of another Bridge . . .' three pages in PRO *State Papers (Dom)* CIX 69 122–3, with a small diagram.
33. Pepys *Diary* 26 October 1664.
34. Journal of the Common Council 5 October 1664, Vol 45 f 423v.
35. *Calendar of State Papers (Dom)* 1663–4 p 670, and *Calendar of State Papers (Venetian)* 1664–6 p 46.
36. Humpherus op cit I p 311.
37. House of Commons 4 April 1671, Grey *Debates* I p 415.
38. CJ 4 April 1671, vol 9 p 230.
39. CJ 21 February and 2 March 1725/6, vol 20 pp 585 and 597.
40. CJ 25 and 30 March, 5 April 1726, vol 20 pp 639, 646 and 653; Journal of the Common Council, vol 57 f 140v.
41. CJ 2 March 1725/6, vol 20 p 597; Fulham Bridge Minutes I p 9.
42. Local Act 12 George I, c 36 (1725) and Act George II, c 18 (1728); Fulham Bridge Minutes I pp 63ff.
43. Fulham Bridge Minutes I p 1.
44. The full list of subscribers to Fulham Bridge is given at the end of Minute Book I and the beginning of Minute Book II.
45. Fulham Bridge Minutes I pp 10–14.
46. Full details are given in the Fulham Bridge Minutes Books which are briefly summarised in C. J. Feret *Fulham Old and New* (1900); see also Manning and Bray *Surrey* (1814) III 287; A. Chasemore *History of the Old Bridge at Fulham and Putney* (1875); VCH *Surrey* IV (1912) pp 78–9.
47. Fulham Bridge Minutes I pp 131–2.
48. Ibid II 2 July 1730.
49. CJ 16 February 1735/6, vol 22 pp 569–7.
50. Fulham Bridge Minutes III pp 190–2, 8 January 1757, and *Gentleman's Magazine* XXVII p 43.
51. LCC *Bridges, Tunnels and Ferries* (1898) p 10.

# 3
# The Decision to Build, 1727–36

It is a curious and perhaps illuminating characteristic of the first half of the eighteenth century, the Augustan Age with which we associate such worldly figures as Marlborough, Walpole, Queen Anne and George I, that its most distinguished architectural monuments were religious and domestic. In an era fundamentally rational and irreligious, the chief glories were not so much Blenheim and Castle Howard and a few buildings at Oxford and Cambridge, but the churches of Wren, Vanburgh, Hawksmoor, Archer and Gibbs. No doubt it could be said that they were worldly too, lacking the spiritual aspiration of the medieval cathedrals, and fitting companions for those other practical beauties of eighteenth-century England, the country houses, the brick farmsteads, and the elegant squares and terraces of the developing towns. Certainly a traveller voyaging up the Thames to Westminster between 1700 and 1740 would have been struck by the multitude of new churches, by the spires of St Anne's Limehouse and Christchurch Spitalfields, and the new towers of St John's and Westminster Abbey. But there seems to have been a reluctance to launch any large-scale construction that might have contributed to a swiftly expanding practical and commercial community. Ambitious projects such as Inigo Jones's Palace of Whitehall and Kent's new Houses of Parliament were discussed, shelved, revived and forgotten. The development of a canal system and the docks had to wait till the end of the century, and the embankment of the Thames itself till half way through the nineteenth. London has ever been a city of lost opportunities and one of its most melancholy casualties is the river Thames. Had it been bordered by Colonel Trench's arcaded embankment (1825) linking the Pool of London to Chelsea, and decorated with such superb temporal features as the Tower, Somerset House, and the proposed palaces of Whitehall and Westminster, the Thames could have been counted among the glories of European civilisations.

However Labelye's Westminster Bridge went a long way towards correcting this bias, though opposition continued long after the first stone had been laid by the Earl of Pembroke in January 1739. After the attempt at the Restoration had been thwarted the bridge enthusiasts lost heart and there was little activity until 1721 when the project was raised again in the shape of two petitions presented to Parliament by the City of Westminster and the Home Counties. The first petition appeared in the name of 'divers Gentlemen, Freeholders and Inhabitants of the City of Westminster and Parts Adjacent' and had obviously been the result of collaboration with the 'divers Gentlemen, Freeholders and Inhabitants of the Counties of Kent, Sussex, Surry and Southampton'. It stated that

> by reason of the great increase of the inhabitants in those parts, an easier Communication is become necessary with the Counties of Surry, Kent, Sussex and other Parts adjacent to Westminster on the other side of the Thames; over which there is no Bridge between London and Kingston, which is very inconvenient, the Carriage of Persons and Provisions by Ferries over the said River being uncertain and hazardous and sometimes impracticable for a long time together; all which inconveniences would be remedied by erecting a Bridge at *Vauxhall* or *Lambeth;* and praying that Leave may be given to bring in a Bill for that Purpose.

The other petition was phrased in similar terms, observing how dangerous it was to cross the river in frosts or tempestuous weather.[1] They were both referred to a committee composed of the Members for Middlesex, Southampton, Surrey, Sussex, Kent and the City of London, and about fifty other MPs including Pelham, Tyrconnel, Walpole, Pulteney and General Wills. 'As to public news', wrote Dr William Stratford, a loquacious but usually reliable gossip, 'we are as in my last. The City continues in a high ferment about the Quarantine Bill and the new Bridge that is to be built between Lambeth and Foxhall. There will be mighty opposition by the City to that Bridge.'[2]

Meanwhile a flood of counter-petitions poured into the House of Commons urging the dangers inherent in such an undertaking. The first to arrive was a petition from John Pond who rented the toll or wheelage of London Bridge and claimed that he had paid a large sum for the lease of which about twenty years remained. He also paid a considerable rent annually towards the repairs of London Bridge and should a new bridge be built at Lambeth or Vauxhall the repairs would have to be reduced, 'to the unavoidable Ruin of the Petitioner and his numerous Family'.[3] This was followed by a petition from the Lord Mayor, Aldermen and Commons of the City of London threatening that such a Bridge if erected will prove inconsistent with and destructive of the Rights, Properties, Privileges and Franchises of the City, will be a great Prejudice to London Bridge and to the Navigation of the River Thames so as to render it dangerous if not impracticable, and will greatly affect the Trade of the City in general and the Properties of many private persons and families in particular . . .[4]

The inhabitants of Southwark said that it would prejudice their trade and increase the number of their poor; the owners and occupiers of houses on London Bridge said that they would be unable to pay their rents and that therefore London Bridge itself would be brought to decay; St Thomas's Hospital said that it would be flooded and drowned together with many poor, sick, lame and diseased people and about 5,000 poor and miserable objects, many of whom were seamen and soldiers; the proprietors of London Bridge Waterworks said that they would be unable to supply the City with water; the Mayor and Recorder of Abingdon in Berkshire said that such a bridge would cause mischief and inconvenience and thousands of people would inevitably be reduced to ruin and want; the tenants of Lord Bolingbroke in Battersea said that they farmed marsh, meadow and garden ground there, that it frequently overflowed in spite of a retaining wall maintained at great expense, and that a bridge would cause the Thames to rise even higher; the bargemen and watermen of the riverside villages from Chelsea to Kingston said that it would be pernicious to navigation and ruin thousands of their families; the lessee of Queenhithe Market, Samuel Cowden, said that the West Country barges would discontinue coming to the market because of the danger of negotiating the bridge; and the inhabitants of Bermondsey, Rotherhithe and Deptford said that, since Dagenham Reach had been blocked up, the tide had frequently caused the river to overflow and another bridge would make this even worse, 'fatal to their possessions and property'.[5] A memorandum by James Brown, a builder who had made sewers in various parts of the town, declared that a bridge would silt up the river and cause the tide to flow higher and higher—'I conceive our river is a pipe which the sea forceth back upon us and as the river is contracted or filled up at bottom, twill occasion higher Tides'—until the sewer system and eventually the river itself would be utterly choked.[6]

While all these petitions were being examined, the Bridge Committee, under its chairman William Pulteney, set to work to examine a number of technical witnesses, among them Sir James Thornhill who, apparently on his own initiative, had measured and bored the river from Lambeth to Peterborough House and had reported that the ground was firm and would make a good foundation. Thornhill's activity in this field is not altogether surprising. He is of course well known for the baroque paintings in St Paul's Cathedral, Blenheim and Greenwich and, especially at this time, at Moor Park and Wimpole. He had become History Painter in Ordinary to George I, Sergeant Painter, Master of the Painter-Stainers' Company, and in 1722 Member of Parliament for his native town of Melcombe Regis. He was an extremely ambitious man and determined to extend his sphere of influence into architecture as well as painting, according to both Vanburgh and Vertue aiming at the post of Surveyor of the King's Works in succession to the incompetent William Benson. In fact Thornhill managed to antagonise the whole architectural profession 'by drawing and designing and demonstrating their ignorance in the Art of Building' to such an extent that early in March 1722 'a mighty mortification' fell upon him and he lost to William Kent the important commission for decorating Kensington Palace.[7] No doubt his survey of the Thames at Westminster formed part of his scheme for the invasion of the architectural world and according to the newspapers he actually produced a plan for the new bridge and presented it to the king.[8] His work certainly helped Pulteney's committee to decide that 'a Bridge where proposed is very practicable and absolutely necessary . . . and may be built and preserved by a reasonable Toll and will be of advantage to the Publick'. The House of Commons agreed and Pulteney and Molyneux, the Prince of Wales's secretary, were ordered to prepare a Bill.[9]

Pulteney's committee had already been at work and acting on the advice of Lord Burlington had commissioned Colen Campbell to design a bridge which, though never further advanced than an austere engraving by Hulsbergh in *Vitruvius Britannicus,* might well have graced the Thames with a delicate example of Palladian architecture forming part of Inigo Jones's Palace of Whitehall, the plans of which were to be revived by Burlington and Kent a few years later.[10]

Campbell's bridge was to have been a 770ft stone structure with five 100ft arches, two 75ft arches and a 650ft watercourse. The plan shows two towers at each end, rusticated niches above the six piers, and a balustrade ornamented with plinths and balls. It was based on Palladio's imaginary bridge, which derived in turn from the Ponte d'Augusto, Caesar Augustus's five-arched bridge at Rimini.[11] Burlington took the precaution of getting two eminent mathematicians, Edmund Halley and Dr Arbuthnot, to check Campbell's design, which was pronounced 'no obstruction to the Navigation of the River nor any Prejudice to the Adjacent Property, which were the pretended Objections of some who were no well-wishers to the Bridge for other Reasons'.[12]

In addition to Thornhill and Campbell, several other architects produced designs. At one time Vanburgh's was the favourite: 'a stone Bridge of nine arches with a large Draw-Bridge in the Centre for the passing of the West Country Barges'.[13] At another time a bridge partly of stone and partly of timber, designed by Ripley, seemed the most likely.[14] The money was to be provided by a syndicate consisting of the Archbishop of Canterbury, a Mr Ward of Hackney and four others, to be repaid at five per cent from the toll and wheelage. Projects were afoot to build squares, streets and inns as far as Tothill Fields, and though opposition was expected the inhabitants of Westminster hoped for great benefits. 'It is computed', said one newspaper,

> that when the new Bridge is built from Lambeth to Westminster, provisions will be sold at least 20 per Cent cheaper than they now are in all the Markets from Temple Bar to Hide Park Corner, as well as in Westminster, there being more than that difference in Price between Southwark, the City Markets and the out Markets by reason the vast quantities of Fish, Fowl, Beef, Mutton, &c. which comes from Kent, Surry and Sussex, are (for want of another Bridge) obliged to come over London Bridge so that the out Markets are served at second hand.[15]

The Bill was read for the second time on 25 January in a House of Commons packed to the galleries and agog with excitement. Sir Constantine Phipps, learned counsel briefed to argue against the Bill, 'spoke a long time, setting forth the Inconveniences that would arise to the City of London and other places by this Project'[16]. Four days later more witnesses were examined, a further hearing was postponed for a week and the Bill was dropped. Its departure was accompanied by one of those slapstick parliamentary scenes by no means confined to the age of Walpole.

> Yesterday Council was heard and Witnesses examined at the Bar of the House of Commons (said a lobby correspondent). The House and Galleries filled with numbers of Strangers ... among them a grave sort of Person who seem'd to be very strenuous for the Bridge (and who by his Habit appeared to be a Projector) was taken in picking the Pocket of Mr Hambleton, a Gentleman belonging to the House of Lords, at the very Door of the House of Commons. The Footmen immediately claim'd him as their Property and conducting him to the Waterside, obliged him to fathom the River, according to an ancient Privilege belonging to those Gentlemen.[17]

The traditional hostility of the City to parliamentary interference with their affairs, especially to the idea of another bridge diverting the traffic from London Bridge and so from the City itself, was reinforced at this time by the almost nation-wide hatred of Sir Robert Walpole. Walpole's financial policy had antagonised the City; his arrogance had alienated both the Court and the Commons; and his reputation as 'the Skreen Master General', or simply 'the Skreen', earned for sheltering the South Sea directors, made him the object of satirical attacks rarely equalled in virulence and crudity. He was of course known to be a sponsor of the Bridge Bill, being a member of Pulteney's committee, and its defeat in the Commons was a symptom of the public odium. However Walpole was a consummate parliamentary tactician, a firm believer in Westminster Bridge, and not easily outwitted. The Bill was shelved for the time being only.

As a sop to its not very tender conscience, the City produced suggestions for a number of 'expedients' intended to relieve traffic congestion. The gateway on London Bridge was to be widened enough to allow two carriages to pass. Street-keepers were to be appointed to keep the traffic moving in the Borough, on Fish Street Hill and on the bridge itself. The country butchers were to remove their stalls out of the main road into St Olave's churchyard. And six new ferry boats were to run at the Lambeth Horseferry 'which the inhabitants of Westminster now seem to want, and for every ferry boat eight horsecloths to cover their passengers' horses if required in wet and cold weather while horses stand still in the ferry boats'.[18] These appear to have been informal suggestions and the only action taken was the publication of the traffic regulations mentioned on page 27.

For the next decade plans for a bridge at Westminster lay dormant. In 1724 General Wade set out for Scotland to build roads '250 miles in length and above 40 Stone Bridges', a feat of military engineering which was crowned in 1735 with William Adam's Palladian bridge over the Tay at Aberfeldy.[19] Maitland briefly mentions a project for a bridge at Westminster in 1725 (possibly a mistake for 1722); and John Bowack, petitioning for employment under the Bridge Commission in 1736, claimed that he had helped the City and Liberty of Westminster to obtain a Bill for a bridge in 1726, though he was probably referring to the Fulham Bridge Act passed in that year.[20]

With the accession of George II the liberal trend of Walpole's commercial policy began to raise the hopes of Westminster again and the king himself said to Lord Egmont 'that a bridge at Lambeth would be convenient, and the clamour the City would raise against it would soon be over, as it was against the bridge at Fulham'.[21] But it was not really until 1733 that the hostility of the City to Walpole's reforms diminished enough for the bridge promoters to feel the time had come. A 'Society of Gentlemen' was formed, meetings were held in the Horn Tavern in New Palace Yard, the river was measured and sounded, and Labelye himself appears on the scene supplying maps and surveys.[22] Thomas Cotton, the magistrate, and Thomas Lediard, later the commissioners' agent, presented a scheme of street improvement. Hawksmoor was instructed to prepare a design. And the City of Westminster set about drawing up a petition to Parliament 'for leave to erect a Bridge over to Lambeth at their own Expence. Plans are already drawn and 'tis said, as the arches are intended to be very large, it will be no Hindrance to the Navigation of the River Thames'.[23]

The 'Society of Gentlemen' was a name coined by Lediard in his pamphlet, *Some Observations on the Scheme offered by Messrs Cotton and Lediard . . . for Opening Convenient and Advantageous Ways and Passages to and from the Bridge,* to describe a group of citizens of Westminster who combined in 1733 to bring up the Bill before Parliament. There were about forty of them including four Justices of the Peace, a sub-dean of Westminster Abbey, a local attorney and a number of substantial citizens living within the parishes of St Margaret and St John the Evangelist.[24] Judging from a list of contributors paying towards the costs of the Bill in May and December 1733, these citizens did not include a group of noblemen and gentry living in the Privy Gardens and around the old Palace of Whitehall, and from whom one might have expected an interest at this time—men such as the Duke of Richmond, the Duke of Dorset, Sir Conyers Darcy, George Treby, Walpole, and the Earl of Pembroke. In fact the only well-known names appearing in the list are those of Nicholas Hawksmoor and Thomas De Veil, the celebrated Court Justice at that time living in Leicester Fields. The obituary notice of Colonel De Veil in the *Gentleman's Magazine* and a *Memoir* appearing in the following year (1747) make no mention of any activity on his part relating to the bridge, but 'he was of an aspiring temper and knew how to bustle through the world', and it may be significant that when the average contribution from members of the Society was four guineas, De Veil and Thomas Cotton were only good for two guineas each and Richard Farwell for one guinea. Admirers of Gobble, Buzzard, Thrasher, Squeezum and the other trading justices of Samuel Butler, Gay, Smollett and Fielding, may draw their own conclusions, but it is only fair to say that the other magistrate, Gideon Hervey, did in fact contribute four guineas. It turned out later that some of the money contributed had been intended as a loan and in 1738 a memorial was presented to the Bridge Commissioners asking for repayment. The affair dragged on for four or five years until Sir Hugh Smithson produced the accounts and a Minute Book, the surviving contributors handed in a petition to the Board, and the 'Society of Gentlemen' was finally dissolved. [25]

A great deal of speculation was caused by the problem of how to raise the £100,000 a stone bridge was calculated to cost. This was a very large sum. The timber bridge at Fulham had cost about £23,000, and Wade's Tay Bridge, built of stone from the nearby quarry at Farrockhill, had cost a mere £4,095. One idea, popular outside Westminster, was that the City and Liberty should raise the money itself and be repaid from the tolls. Another was for life annuities to be granted at ten years' purchase with the tolls as security. As the annuitants died off so the tolls would decrease and the bridge eventually become free. A third proposal was for money to be raised by subscription, as was done for Fulham Bridge. Neither a lottery nor a government grant was considered at this stage not was it ever intended to tax Westminster itself. [26]

It is pleasant to think of Nicholas Hawksmoor, by this time an invalid of over seventy, tormented with gout and almost overwhelmed by victimisation of the most unscrupulous kind in his professional career, returning to a major project in London where he was at that very moment engaged on the towers of the Abbey. His bridge, of which the first plans were prepared in the autumn of 1734, was to have had seven arches, the middle arch of 120 feet, the next two of 100 feet, and the remaining five of eighty feet each. [27] These were modified to nine arches, and a stone model was shown to the Prince of Wales and the House of Commons. [28] It was to be built at the Horseferry where the river was 830 feet wide with a hard gravel bottom. A drawing made in March 1736, the month of Hawksmoor's death, shows the middle arch of a magnificent bridge starting from New Palace Yard and decorated with the royal arms, classical niches above the piers, a balustrade surmounted with urns, and at intervals pedestals for royal statues supported by lions and unicorns. This was probably a fantasy. Hawksmoor's published design appeared in his pamphlet and was subdued in character. [29]

Hawksmoor, with the help of Charles Labelye, calculated that with 660 feet of waterway between the arches, the rise and fall at the bridge would be little more than four and never more than five inches, whereas one of the great hazards of navigating London Bridge, where the river was 900 feet wide and crammed through a waterway of less than 200 feet, was the awesome cataract at spring tides of something like six feet roaring between the sterlings. Labelye himself said that Hawksmoor's bridge would cause so small a fall that it would never 'hinder any Vessel or Boat from rowing through, even at the swiftest Tides'.[30] John James, reviewing three of the designs submitted (those of Price, Hawksmoor and Batty Langley), was not so sure. Some observers, he pointed out, calculated that at low tide the water level was actually higher at London Bridge than at Lambeth and a new bridge therefore might have the very opposite effect and cause considerable flooding. He also decided that Hawksmoor's bridge was wider than necessary. It was based on the bridge over the Loire at Blois, which was over 100 feet longer than Hawksmoor's at Westminster and ten feet narrower. However it was 'designed with good Judgement', he conceded; 'the Arches are of the largest but then he has made them semicircular and taken their springing at the Low Water Mark which adds very much to the Strength of the Legs that are to sustain them'.[31]

Another entrant at this stage was John Price, one of the contesters for Fulham Bridge and the architect of St George's Church, Southwark, completed that year. His plans were presented to the House of Commons but were never seriously considered.[32] Price's bridge was also to be built at the Horseferry and to consist of nine arches, but according to James, the piers were too high to bear the thrust; and furthermore it was only forty feet wide, reducing the roadway to twenty feet, which would certainly have caused traffic blocks. A great deal more shocking, however, was James's accusation that Price had quoted long extracts from Palladio and Gautier without acknowledgement to either—a particular discourtesy to Gautier, who 'has deserved very highly of all who have either occasion or desire to be instructed in the Business of building Bridges'.[33]

Early in 1735 the *London Daily Post* had reported that the scheme was to be laid aside and no petition presented that session. But in February a series of weekly meetings in the Horn Tavern urged on the project which this time was not to be allowed to founder. A bridge at Westminster, it was decided, was 'a Thing so necessary and convenient, not only to all the Inhabitants of the said City and Liberty but greatly beneficial to the Counties of Middlesex, Surry, Sussex and Kent; and likewise to a great part of London'.[34] The Horn Tavern meetings were encouraged by the two Members for Westminster, Lord Sundon and Admiral Wager. Wager introduced the committee to Walpole himself, a keen supporter of the idea, who promised his official approval at the proper time. A memorial presented several years later to the Bridge Commissioners by the survivors of the Society of Gentlemen, drew attention to their activities at this time in the form of an account for £260 for measuring and boring the river, hiring barges, surveying the approaches, making maps and sections, and preparing the Bill.[35]

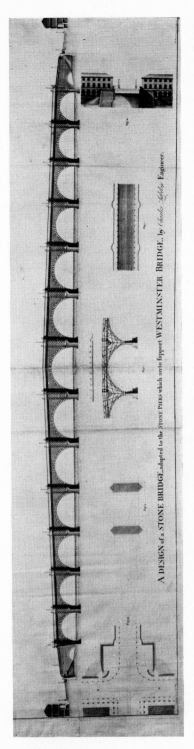

A DESIGN of a STONE BRIDGE, adapted to the STONE PIERS which are to support WESTMINSTER BRIDGE, by *Charles Labelye* Engineer.

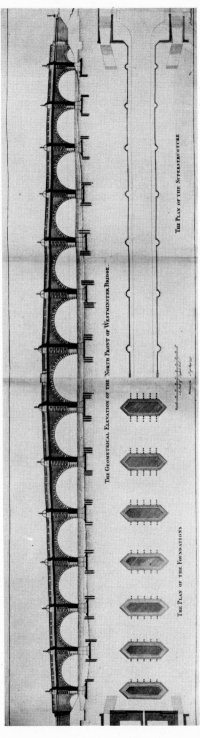

THE GEOMETRICAL ELEVATION OF THE NORTH FRONT OF WESTMINSTER BRIDGE.

THE PLAN OF THE FOUNDATIONS

THE PLAN OF THE SUPERSTRUCTURE

5. Labelye's first design for Westminster Bridge, 1736, with a note of the 1743 alterations. (*Engraving by Fourdrinier in the House of Lords. Photograph by courtesy of the Department of the Environment*)

6. Labelye's final design for Westminster Bridge, 1739. (*Engraving by Fourdrinier in Sir John Soane's Museum. Photograph by R. B. Fleming*)

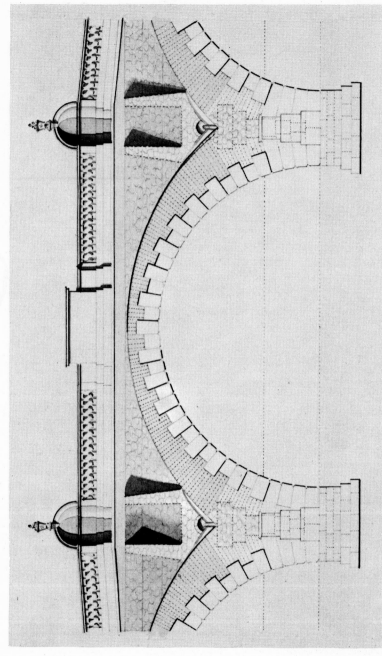

7. Section of Labelye's final design. (*Drawing by Thomas Gayfere in the Institution of Civil Engineers. Photograph by R. B. Fleming*)

In January 1736 'a great many of the Burgesses and other Inhabitants of the City and Liberty of Westminster met in Fish Yard near the Abbey and resolved on a Petition to Parliament for a Bridge across the Thames from Lambeth to Westminster Horse-Ferry' and the Burgess Court of Westminster agreed that every burgess should subscribe two guineas towards the costs of the Bill.[36] The petition was presented to the House of Commons on 4 February by Admiral Wager who had been a champion of a new bridge for many years—in fact since 1725, when he had been one of the most active members of the commission for Fulham Bridge. The petition declared

> that the said City and Liberty hath for many years past greatly increased in Number of Buildings and Inhabitants and is now become very Populous; and that it will be advantageous, not only to the said City but to many others of His Majesty's Subjects and to the Publick in general, to have a Bridge erected across the River Thames from the Horseferry at Lambeth, for their more convenient Communications with the several Counties of Surry, Kent and Sussex...

Thereupon the Commons appointed a committee of about a hundred members (including Walpole, Wager and Wade) and ordered it to examine the matter thoroughly and report back to the House.[37]

While the committee was at work, the usual opposition arose from the City, the watermen and various interests up and down the river. But a new and unforeseen obstacle appeared from the heart of Westminster itself, where a group of citizens declared that 'if a Bridge be built it will cause the Water to flow more abundantly and probably lay most of the Ground Floors under Water'.[38] This gloomy prognostication was dramatically fulfilled exactly a week later while the Bridge Bill was being debated in the House of Commons. 'The River overflow'd its Banks by reason of a strong Spring Tide, the Water was higher than ever known before and rose above two feet in Westminster Hall where the Courts being sitting, the Judges &c. were oblig'd to be carry'd out. The Water came into all the Cellars and Ground floor rooms near the River on both sides and flow'd through the streets of Wapping and Southwark . . .'[39] The damage was extensive further down the river, especially at Limehouse and Rotherhithe, where the inhabitants were in great distress. In the marshes near Rochester, sheep and cattle were drowned by the hundred. Floods of this severity were unusual and the opposition was quick to point out this one as an omen not to be ignored.

The committee's report was delivered to the House of Commons on 16

February and read out by Lord Sundon, the Member for Westminster.[40] It was illustrated with six models—some of wood, some of stone—displayed in the anterooms outside the House. A map of the river showed five possible sites for a bridge: the Horseferry, New Palace Yard, the Slaughter House, Derby Court and Whitehall (Plate 1). Below Whitehall the river curved eastward and most people agreed that the set of the tide across the bend would both endanger the navigation and undermine the foundations. Sundon reported to the House that the committee had examined various witnesses, the most important being Charles Labelye (here spelt La Belie), Dr Desaguliers and Captain Knowles. Labelye began by saying that he had lived in England for sixteen years and during that time had noticed a great increase in the size of Westminster where a bridge would be a convenience and a means of lowering prices in the local markets. He favoured the Horseferry site where he had bored the bed of the river across to Lambeth, finding it to be mostly of hard gravel and a good foundation for a stone bridge. Further down-stream it became muddier and at Derby Court the drill could be drawn out by four men, whereas at the Horseferry it stuck in the gravel and six men could not move it. Further down still, towards the bend in the river, the ground became even softer, until at Whitehall it was quite unsuitable for a bridge. Moreover the big West Country barges, some 120 feet long, would have a difficult passage coming up on the flood tide because of the oblique set through the arches. Labelye at this stage scarcely considered himself to be a likely candidate, still supporting Hawksmoor with whom he had worked for the last year. His evidence was confirmed by two surveyors, Hinton and Pattison, who had both been present at the borings.

Dr J. T. Desaguliers had also given evidence, agreeing with Labelye that the nearer Westminster they drilled the more mud they found, though he thought the river-bed itself would make a good foundation. Desaguliers, a Huguenot refugee after the Revocation, was consulted several times on problems connected with Westminster Bridge. He was an extremely distinguished mathematician and natural philosopher, living in Channel Row near Derby Court and taking a great interest in Westminster affairs. His book, *A Course of Experimental Philosophy,* had been published two years before this, and he was responsible for an improved ventilator in the attic over the House of Commons. He supported Hawksmoor's plan for a stone bridge at the Horseferry site and was a firm opponent of a timber bridge.

Other witnesses examined were Mr Wilkey (the master of a hoy), Mr Price (a barge owner), and Mr Vaston (a sugar baker), all of whom preferred the Horseferry because of the direct current and the depth of water giving river traffic a fair passage clear of shoals and tidal hazards. Wilkey in particular seems to have taken a lot of trouble, testing the flow of the current with float buoys at various stages of the tide and observing the exact set at all five of the proposed sites. As a professional waterman he had no doubt that the Horseferry was the best place, though it is probable that both he and Price were influenced by the watermen's calculation that their Sunday ferries at Lambeth were worth £1,000 a year and that compensation would be granted in proportion.

The last witness to be examined was a distinguished naval engineer, Captain Charles Knowles. Knowles had served in the Mediterranean with Wager who was aware of his abilities and no doubt had co-opted him as a witness. He took a different view from the others, preferring Whitehall as a site. He held that Whitehall was more central than the Horseferry and had more space than New Palace Yard, where he believed traffic congestion would impede Members of Parliament going to the House. Moreover at Lambeth the street leading to the Kent roads was narrow, a great deal of property would have to be bought including the Archbishop's interest in the ferry, and the banks would have to be artificially raised.

The subject provided an endless source of argument both in Westminster and in London as a whole. Knowles had a supporter who wrote a long letter to *The Post Boy* advocating Whitehall or even a site as far downstream as Scotland Yard. As the problem 'now engrosses a great part of the Conversation in this City and that Posterity may see some of the Reasoning of the present Age' his letter was republished in *The Political State of Great Britain*.[41] *The Post Boy* correspondent argued at some length that the most proper place was not necessarily the narrowest part of the river, pointing out that however wide the arches were built, the piers were bound to act in some degree as a dam and cause a fall in the water level.

> And if a Bridge was to be built at the Horseferry, not capable of letting more Water pass in the same time that *London* Bridge is now able to pass, the Fall would be near as much at *Lambeth* Bridge as it is now at *London* Bridge, and to have passed against the Stream would have been impracticable. Hence the Folly of making a Bridge at the narrowest Part of a navigable River, that is to consist of more than one Arch.

In his opinion the best place for a bridge would be either in the Strand from the end of Southampton Street or from Charing Cross or Scotland Yard where the river was at least 300 feet broader than at the Horseferry. The objection put forward that the river curved near Whitehall and that a bridge built there would cause the formation of shoals he dismissed as irrelevant. Provided the point chosen was in the centre of the curve so that the tides setting on either bank were equidistant, the curve need be no obstacle. In order to determine the exact spot, he said, 'in a calm Day take a long West Country Barge to the place proposed, in the Middle of the Stream, so that when she rides without any Sheer upon the Stream, the place you pitch upon to build your Bridge from should be exactly on her Beam'. He then proposed that careful tidal observations should be made, stakes driven in where the piers were to be built and the longest and heaviest-laden barges taken through to decide whether they could pass without touching. 'I cannot but be of the opinion', he concluded, 'that *Whitehall* or *Scotland Yard* are far preferable to any Place or Part of the River that is not so wide.' The letter was unsigned but could conceivably have been from Knowles himself.

The report recommended to the House that a bridge ought certainly to be built; but, contrary to the opinions of the witnesses and with no indication of how such a decision had been arrived at, it suggested that the most useful position would be from New Palace Yard. The choice of New Palace Yard was not altogether surprising even though it presented many difficulties, especially the acquisition of property for the approaches on either side. It was the middle site of the five proposed by the bridge promoters, a compromise between the narrow stretch at Lambeth where the ferry already existed, and the treacherous oblique currents below Whitehall; and—most blessed virtue of all to men endlessly frustrated by the idiosyncrasies of the Thames watermen—it would deposit Members of Parliament at the very doors of the Palace of Westminster itself. Without more ado the House agreed and ordered Lord Sundon, Admiral Wager, General Wade, Colonel Bladen and John Crosse, all citizens of Westminster, to prepare a Bill. Unfortunately for the historian, a motion to print the report was negatived.[42]

The expected opposition flared up immediately but less violently than in 1722. Hawksmoor's hope 'that the City of *London* of such mighty Commerce . . . will not only not oppose but readily promote another Bridge, for a better Communication of Trade and joint advantage of both *London* and *Westminster*', proved an illusion, as he doubtless anticipated.[43] The Court of Common Council invoked the familiar charter of Henry VI to prove that the soil of the Thames had been granted to the City, and sent up a petition to the House of Lords begging leave humbly to represent that the erecting a Bridge at Westminster, but more especially so low on the River as New Palace Yard, will prove destructive of several Rights, properties, privileges and Franchises of this City; will be a great prejudice to the Navigation of the River of Thames, so as to render it dangerous if not impracticable; and the same will greatly affect the Trade of this City in general.[44]

The proprietors of Fulham Bridge expressed themselves anxious not to obstruct a 'Design of so General a Benefit' but nevertheless petitioned the Commons for relief on the grounds that, should the new Westminster Bridge be built, their property would be prejudiced and, if it were to be toll-free, 'very near if not totally destroyed'.[45] The Company of Watermen and Lightermen lost no time in defending their traditional cause and claimed that the new bridge would obstruct navigation,

greatly lessening if not totally destroying several ferries between Vauxhall and the Temple . . . by which means the Old and Decrepit will become chargeable upon their respective Parishes, and the Young and Able must be obliged to seek their Bread in Foreign Parts . . .

The lightermen in charge of coal barges sent up a similar petition as did the inhabitants of Southwark who foresaw frequent floods and being overwhelmed not only by the river but also by watermen thrown out of work, 'who are Parishioners and inhabitants and from want of Employment must become a Burden'.[46]

However, on this occasion a path for the Bill had been more carefully prepared. The meetings in the Horn Tavern and Fish Yard, the articles and letters in the newspapers, the spectacular growth of London over the last few years and the influx of immigrants both from England itself and from France and the Low Countries, had created a climate of opinion different from that which had prevailed at the time of the previous attempt in 1722. Most people were determined to have a bridge and intolerant of opposition. The great capital cities abroad, and even the provincial towns of impoverished countries such as Spain, were far ahead of London. 'If the provident *Dutch* were Masters of this River, they would have at least five or six Bridges between *Billingsgate* and *Westminster*', said Caleb D'Anvers in *The Craftsman*.[47] It was monstrous that so great a maritime and trading nation should possess only one antediluvian bridge at the entrance of a navigable river the size of the Thames. The Commons lost no time in bringing in the Bill. Lord Sundon presented it on 23 February, when it was received and read for the first time. Wager informed the

House that the king had consented to open a passage through part of the Privy Gardens, should this be necessary. It was read a second time on 27 February and studied by the whole House in committee on 15 March when Lord Sundon reported the committee findings; it was ordered to be engrossed on 19 March. Several petitions were read and counsel heard again on the 29th; on 31 March a last fling by the opposition to add a rider that a jury be appointed to examine the dangers to navigation was immediately squashed, as was another clause for the relief of decayed watermen or their widows. The Bill was passed by 117 votes to 12 and Sundon was desired to carry it to the Lords for their concurrence. At the Bar of the House of Lords, Dr Desaguliers, Dr Halley, Captain Knowles, and the master of the Mathematics School at Christ Church, Mr Mitchell, were once again examined and made to demonstrate with models that the bridge would in no way damage navigation.[48] In spite of a petition from the Court of Common Council, the House of Lords agreed to the Bill, with amendments relating to compensation to the Archbishop of Canterbury for his interest in the Lambeth Horseferry, to the poor watermen of St Margaret's and St John the Evangelist's in Westminster, and to the Company of Wherrymen, Watermen and Lightermen between Gravesend and Windsor.[49] On 5 May the Commons considered the Lords' amendments and after the journals had been read for the various dates when the Commons had disagreed with the Lords, the amendments were agreed. On 20 May 1736 the Bill received the Royal Assent.[50] Parliament was prorogued and the way was at last clear for the bridge builders to begin.

## NOTES TO CHAPTER 3

1. CJ 15 December 1721, vol 19 p 694.
2. Dr Stratford's letter to Edward Harley 29 December 1721, *Hist MSS Comm. Portland* 7 pp 285 and 312.
3. CJ 9 January 1721/2, vol 19 p 726.
4. CJ 18 January ib. The petition was presented by the sheriffs at the Bar of the House as a result of the work of a committee appointed by the Court of Common Council 20 December 1721 (Journal vol 57, f 93 and 95v).
5. CJ December and January 1721/2, vol 19. The objections were embodied in a broadsheet, *Reasons against Building a Bridge* . . . (Bodleian Library, Gough Westminster 22), and in a pamphlet, *Reasons against building a Bridge from Lambeth to Westminster showing the Inconveniences of the same to the City of London and the Borough of Southwark* (1722).
6. Guildhall MSS 119.8, 19 January 1721/2.
7. *Vertue Notebooks* I pp 100–1 and III p 55. For Thornhill as architect and

painter see Colvin *Dictionary* pp 610–12 and Edward Croft Murray *Decorative Painting in England* I (1962) pp 265–6.

8. *Daily Journal* 17 January 1721/2. Drawing in RIBA.

9. CJ 11 January 1721/2, vol 19 p 708.

10. *Vitruvius Britannicus* III (1725) plate 56; H. E. Sutchbury *The Architecture of Colen Campbell* (1967) pp 60–1.

11. Palladio *The Four Books of Architecture* (1570) III p 14.

12. *Vitruvius Britannicus* III p 10.

13. *Daily Journal* 16 January, *London Journal* 20 January 1721/2.

14. *Weekly Journal* 20 January 1721/2.

15. *Daily Journal* 26 January 1721/2.

16. Ibid.

17. *Daily Journal* 30 January 1721/2.

18. 'Expedients to hinder a Bridge at Westminster', Guildhall MS 11918 (undated).

19. J. B. Salmond *Wade in Scotland* (1938) pp 141–3. *Vitruvius Scoticus* for Adam's plans for the bridge.

20. Maitland *History of London* (1775–6 edition) I p 48.

21. *Hist MSS Comm*. Egmont Diary I p 33, 7 February 1729/30.

22. Labelye *Description* p 69; Samuel Price was paid £26 in March 1739 'for services, Barge hire & Trouble in boring the River by order of the Gent. who met weekly in 1734 & 1735' (Sub Accounts I p 27).

23. *Read's Weekly Journal* and *The Weekly Oracle* 14 December 1734; Lediard *Observations* (1738) pp 10–11.

24. The four Justices were De Veil, Cotton, Farwell and Gideon Hervey; the sub-dean was Dr Kendrick, whose chief claim to fame was a funeral oration 'The Man without Guile', preached in St Margaret's in 1753; the attorney was Mark Frecker, who later helped the commissioners on a number of legal points.

25. Bridge Minutes 2 February, 4 May and 8 June 1743.

26. *London Daily Post* 24 December and *Read's Weekly Journal* 28 December 1734.

27. Ibid. 'We are all for a Stone Bridge', said Hawksmoor, 'and we hope No body will oppose it, unless some wooden-headed carpenter'.

28. *London Evening Post* quoted in *Grub Street Journal* 9 January 1734/5.

29. Engraved by Toms as a plate in Hawksmoor's *A Short Historical Account of London Bridge with a Proposition for a New Stone Bridge at Westminster* (1736). The drawing (in the RIBA library) is one of the last from the hand of this great architect.

30. Hawksmoor op cit p 19.

31. James *Short Review* p 19.

32. John Price *Some Considerations humbly offered to the Hon Members of the House of Commons for Building a Stone-Bridge over the River Thames from Westminster to Lambeth* (1735), two copies in the Gough Collection, Bodleian Library, one with amended measurements and a print showing three arches and four piers with niche and œil-de-bœuf, dated 25 February 1734/5.

33. James op cit p 13.

34. *London Daily Post* 12 and 17 February 1734/5.

35. Bridge Minutes 29 November 1738 and 6 April 1739. It was signed by William Morris, John Langley, William Cowper, Thomas Scott, William Miller, Robert Nicholls, Edmund Burton, Richard Farwell, Peter Elers, Samuel Saville, Samuel Gell, Gilbert Marshall, William Paine, John Mackreth, Thomas De Veil, H. Moody, Thomas Rea, John Broughton, Alexander Wilson, Edmund Ball and William Wright.
36. Burgess Court Minutes 20 January 1735/6, p 257.
37. CJ 4 February 1735/6, vol 22 p 547.
38. *Daily Journal* 9 February 1735/6.
39. *Daily Post, Daily Journal* 17 February, *Read's Weekly Journal* 21 February 1735/6; *Gentleman's Magazine* VI (1736) p 109.
40. CJ 16 February 1735/6, vol 22.
41. *The Post Boy* 19 February 1735/6, reprinted in *Political State* 51 pp 252–6; see also *Political State* 54 September 1737, pp 261–2.
42. CJ 16 February 1735/6, vol 22 p 571.
43. Hawksmoor op cit p 11.
44. Journal of the Court of Common Council 26 February 1735/6, vol 57 f 369v, and 5 April 1736, vol 57 f 374v.
45. Fulham Bridge Minutes 3 March 1735/6, vol 2 p 195.
46. CJ 22, 25 and 29 March 1736, vol 22 pp 642, 652, 659, 667–8.
47. *The Craftsman* 13 September 1735.
48. *Daily Journal* 15 April 1736. Several other experts attended, whose names are not mentioned but probably included Labelye and Ripley. Hawksmoor had died a few weeks earlier.
49. *Lords Journal* 19 April 1736, vol 24 pp 649–50, and CJ 20 April 1736, vol 22 pp 690–1.
50. CJ 5 and 20 May 1736, vol 22 pp 705–6 and 718.

# 4
# After the Act of Parliament, 1736-7

The cause which had been fought over with such bitter feelings for the past seventy years at last became law in May 1736. It was called 'An Act for Building a Bridge across the River Thames, from the New Palace Yard in the City of Westminster to the opposite Shore in the County of Surrey'.[1] It began by stating firmly that a bridge would be 'advantageous not only to the City of Westminster but to many other of His Majesty's Subjects and to the Publick in general', and then passed on to nominate about 175 commissioners including the Archbishop of Canterbury, the Lord Chancellor, the Lord President, a galaxy of dukes, marquesses, earls and barons about 100 Members of Parliament, and the Surveyor of the King's Works, Richard Arundell.[2] They were ordered to meet on 22 June in the Jerusalem Chamber in Westminster Abbey and thereafter at their own convenience, nine to form a quorum. When the bridge was finished its architect paid the commissioners a handsome tribute: 'notwithstanding their great Trouble, Care and wearisome Attendance, in the Discharge of the several important Trusts reposed in them by the Legislature, [they have had] absolutely no kind of Salaries, Perquisites, Fees, Rewards or Considerations whatsoever except (as a Nobleman among them nobly expresses it) *the Honour of doing what was thought impossible.*'[3]

The Act gave them power to decide on the materials to be used, to make contracts, to treat with landowners and to pull down streets and houses at the points of access. It forbade them to allow houses to be built on the bridge, the results of a lesson painfully learned from the overcrowding on London Bridge. Seven hundred and sixty feet of waterway, or about two thirds of the width of the river at New Palace Yard, was to be left clear under the bridge. With an eye to the future machinations of the watermen and other traditional opponents, it declared that should anyone 'wilfully and maliciously blow up, pull down or destroy' the bridge, or endanger the lives of its passengers, they should be treated as felons and suffer death without benefit of clergy.

The commisioners were then authorised to treat with landowners in general and with the Archbishop of Canterbury for losses estimated for the Lambeth Horseferry; and with the poor watermen of Westminster in respect of Sunday ferries. The Act then arrived at the vital subject of how funds should be raised to finance the project. The target was £625,000 to come from a state lottery of 125,000 tickets of £5 each. The arrangements were to be in the hands of a group of 'managers' who were to be paid £2,000 for the responsibility. The prize money was to amount to £525,000 and the remaining £100,000 to be set aside for the purchase of land and houses, making the approaches, widening the streets, and building the bridge itself and keeping it in repair.

Sections 13 to 34 of the Act dealt with the lottery details: how the tickets should be numbered, how the contributors or 'adventurers' should sign, when the books should be delivered to the managers, and so on. The counterfoils were to be carefully rolled up and fastened with silk, one column to be deposited in a box marked (A) which was to be locked with seven different locks and sealed by seven of the managers, the other column to remain in the books and be set aside as a check should any mistake or fraud be discovered. Another set of books was to be made containing a similar sequence of 125,000 numbered tickets, the outer column inscribed 'The Bearer of this Ticket, in case it be drawn a Prize, is intitled to the Prize so drawn, within Forty Days after the Drawing is finished'. Of these, 30,613 were to be fortunate tickets. There was to be a first prize of £20,000, two second prizes of £10,000 each, three of £5,000, ten of £3,000, forty of £1,000, sixty of £500, and so on down to 28,800 £10 prizes. In addition the owners of the first and last tickets to be drawn were to win £1,000 each. The tickets were to be rolled up like the counterfoils, carefully fastened with silk and deposited in a box marked (B) to be locked and sealed in the same way as box (A). The draw was scheduled for 23 April 1737 and was to take place in public, ten days' notice being posted in the *London Gazette*. Counterfeiting or forgery of tickets was to be judged a felony with the penalty of death without benefit of clergy, and heavy penalties could be incurred by an officer diverting or misapplying the money. Prizes were to be tax-free; the Bank of England was to be given a proper allowance for its 'labour, pains and service' in receiving the money and keeping it in safe custody; the cashiers were to be properly paid; and the managers were to have £100 each to be paid ten days after the draw and a further £100 on or before 10 October 1738 'for their Labour, Pains and Attendance in delivering out Certificates in Exchange for Fortunate Tickets'.[4]

Finally the Act directed that quarterly accounts should be rendered to

Parliament and that the bridge should rate extra-parochial—which meant that it was to be maintained by the commissioners and not to become a burden on the local rates, provided that it did not become, at some future date, a county bridge subject for maintenance and repair to Surrey and Middlesex. A proviso was inserted protecting the jurisdiction of the Lord Mayor and the City of London.

The decision to finance Westminster Bridge by means of a lottery was a historic one in the sense that this was the first occasion in England that a lottery had been used for this exact purpose. It was generally realised that a bridge was necessary by this time, but most people expected it to be financed either by private enterprise or semi-officially by the Home Counties. The parliamentary resolution to make it a state responsibility came as a surprise. However, state lotteries to finance public works had been used frequently since Queen Elizabeth's lottery of 1569, 'a verie rich Lotterie Generall', which aimed at paying for 'the Reparation of the Havens and Strength of the Realme, and towards such other publique good Workes'.[5] The Elizabethan lottery was a failure, the organisers finding themselves unable to cajole a sceptical public into subscribing more than a fraction of the £200,000 target. But that did not deter the launching of a whole sequence of state lotteries throughout the reigns of James I and Charles I, many of them extremely successful. The Virginian lotteries raised funds to help the struggling settlers in the first American colony of Virginia, and the London aqueduct lotteries were arranged to finance a scheme to improve London's water supply by building an aqueduct from the springs at Hoddesden in Hertfordshire. During the Interregnum lotteries were stifled, as might be expected, but they appeared again at the Restoration, Charles II supporting a number of public and private enterprises including the Thames waterworks lottery, a royal fishing lottery to strengthen the fishing industry, and a series of lotteries run by 'our trusty and well-beloved cosmographer' John Ogilby to finance the publication of his *Britannia* in 1675. But very shortly after the beginning of the reign lotteries began to fall into disrepute, becoming notorious for corruption and double-dealing and leading to the impoverishment and misery of the unlucky adventurers and their families. By the end of the seventeenth century it became clear that private lotteries must cease. As it was equally clear, however, that lotteries in general could afford an easy way to increase the national revenue, legislation was put in hand to monopolise the whole system for the benefit of the nation. Both William III and Queen Anne used them to help carry on the war against France and to assist the Exchequer in general, though for the first decade of the eighteenth century they were suppressed, mainly because

> several evil-disposed persons, for divers years past, have set up many mischievous and unlawful games, called *lotteries,* not only in the Cities of *London* and *Westminster,* and in the suburbs thereof and places adjoining, but in most of the eminent towns in *England* and *Wales* . . . to the utter ruin and improverishment of many families.[6]

After this virtuous interval they were in regular use from 1710 onwards until the last Act authorising a state lottery was passed in 1823. Since then lotteries, though technically illegal in England, have flourished as ballots, sweepstakes, Tombola, Premium Bonds, Bingo and other thin disguises.

Most of the eighteenth-century state lotteries, especially in the early part of the period, were highly successful and often over-subscribed. But for various reasons the Westminster Bridge lotteries, of which there were five between 1736 and 1741, were failures and the bridge had to be subsidised by annual parliamentary grants.

The Act itself proved inadequate in other ways and had to be amended annually during the following eight years and on two occasions in the 1750s[7]. Apart from the question of raising money, the chief difficulty was the complicated business of buying up land on both sides in order to build the approaches, and the commissioners were forced to apply to Parliament every year for a new Act increasing their powers to negotiate with landowners. Many of these were tractable and public-spirited, others avaricious and difficult. In Westminster, where the streets were narrow and houses dilapidated and often labouring under sub-leases and mortgages, the commissioners found the Act hopelessly ineffective and had to empanel juries in order to help them to come to decisions. On the Surrey side it was hardly less complicated, the approach running over Lambeth Marsh and through a maze of market gardens and orchards, the actual ownership proving difficult to determine. Moreover much of the land there belonged to the City, which, though aware that a new bridge had become inevitable, still tried to protect its interests with a rearguard action. Official bodies such as the Dean and Chapter of Westminster Abbey and the Archbishop of Canterbury were also interested landowners, bargaining to the last penny in the name of their successors. Under the Act of 1736 the commissioners were virtually impotent and compelled repeatedly to appeal to Parliament for more and more statutory power.

The first meeting of the Bridge Commissioners took place on 22 June 1736 in the Jerusalem Chamber as the Act had ordered. Lord Sundon was in the Chair and fifty-four commissioners were present.[8] Sir Joseph Ayloffe was appointed Clerk and Nathaniel Blackerby Treasurer, 'upon his giving good and sufficient security in the penalty of Ten Thousand pounds for his due executing that Office'. The Bank of England cashiers, James Collier and John Waite, were told that should a lottery take place the commissioners would allow the Bank £2,000 for its care and trouble. They then examined the models which had been exhibited in the House of Commons earlier in the year, and adjourned for a month. [9]

The second meeting was held on 23 July, also in the Jerusalem Chamber, where the commissioners continued to meet until August 1737 when the Bridge Office in Duke Street was ready. A number of new names appeared among the commissioners present: Thomas Archer (MP for Warwick, not the architect of St John's Smith Square), David Papillon and Thomas Clutterbuck, the last two becoming fairly regular attendants and Papillon a member of the Works Committee the next year. After the minutes of the previous meeting had been read and confirmed, the Bank of England cashiers were called in and reported that the list of subscribers to the bridge lottery showed that 24,631 tickets had been taken and £14,822 already paid into the bank. This was encouraging, the optimum being 125,000 five-pound tickets. The clerk, Sir Joseph Ayloffe, then produced an instrument in writing appointing himself to the office and this was signed by the commissioners present.

Ayloffe had been educated at Westminster School and now lived in Westminster. Fairly early in life he became interested in antiquities and at the age of twenty-two he was elected Fellow of the Royal Society and of the Society of Antiquaries to which he contributed several papers. He published a quantity of historical and antiquarian works of which the best-known is *The Calendars of Ancient Charters . . . in the Tower of London*. His appointment as clerk to the Bridge Commission was doubtless due to nepotism or graft, as was the custom of the time, but he certainly attended meetings regularly, even when the commissioners failed to turn up. He appears throughout the construction of the bridge to have managed its affairs efficiently, though it must be admitted that most of the hard work was done by the commission's surveyor and agent, Thomas Lediard.

A remarkable petition, dated 22 June 1736, was then presented to the board by Andrews Jelfe and Samuel Tufnell (masons), Thomas Phillips and John Willis (carpenters), Joseph Pattison (blacksmith), and George Devall (plumber), complaining that with great concern they have heard it is apprehended that there are not people in England capable of Building the said Bridge. They therefore most humbly pray that they may be indulged to build one Pier of solid stone from the bottom of the River to three feet above High Water Mark at their own hazard and expense, and of such dimensions as they shall be directed, and in that part of the River thought most difficult.

They stipulated that proper overseers should be appointed to inspect the materials and keep accounts so that the commissioners would be able to judge how to contract for the whole bridge. This offer showed some acumen by the building industry and was successful in influencing the commission, at least as far as concerned Jelfe and Tufnell who were awarded the masonry contract.[10] The reflection on the industry's professional ability was probably not taken too seriously. Considering its accomplishments over the past fifty years, culminating in a tour de force of the magnitude of St Paul's Cathedral, architects and builders had no need to be unnecessarily modest. However there is no doubt that schemes were afloat, in Paris at least, to find an architect willing to build the bridge. The ambassador himself, Lord Waldegrave, was probably involved, and a French government contractor, Pierre Lambot, actually offered to build the bridge, at £12,000 a year. Captain Knowles was the moving spirit of this enterprise. He was in touch with a military engineer, the Chevalier Hermand, plying him with the various pamphlets appearing at this time, 'so that you can see for yourself the ignorance and vanity of our architects'. Even the commissioners, says Knowles in one letter, 'are convinced of the inadequacy of our people and of the need to have someone from Paris to take over the building'.[11]

A number of other petitions were laid before the board at this meeting, including one from John Bowack, who claimed that he had helped the City and Liberty to obtain a Bill for a bridge in 1726, and that in December 1734 he had organised the Horn Tavern meetings which had finally steered the Bill through Parliament. Bowack petitioned for employment under the commission and was appointed Assistant Clerk under Sir Joseph Ayloffe at a salary of £100 a year. Nathaniel Blackerby then produced as securities for his office of Treasurer Robert Mann of Chelsea (£3,000), Richard Farwell of Old Palace Yard (£1,500) and Benjamin Wilson of Highgate (£1,500), and the clerk was ordered to inquire into their circumstances and report at the next meeting.

Blackerby was a distinguished Westminster character living at Parliament Stairs in two Old Palace Yard houses which later became the Bridge Office. He had formerly been treasurer to the Commission on Fifty

Churches, had married Hawksmoor's widowed daughter Helen, and since 1735 had been housekeeper of the Palace of Westminster. He was a Deputy Lieutenant of the Westminster militia and figured frequently at the Westminster sessions where he was a justice of the Court of Common Pleas, becoming chairman in 1737. He was made treasurer of the Free Masons in 1738; and in April 1742 he died of pleurisy, in debt to the Bridge Commission due to the bankruptcy of one of his securities, redeemed with some difficulty by his son Benjamin Blackerby.[12]

During the summer of 1736 there occurred two developments of interest. The first was the entry of Batty Langley into the lists with a design for a bridge accompanied by a pamphlet. The second was the public burning of the Bridge Act in Westminster Hall while the courts were actually sitting. Batty Langley's appearance caused no surprise. He was an irrepressible jack-of-all-trades and author of a number of treatises on building and design. The previous year he had been employed by Nathaniel Blackerby to erect at his Parliament Stairs house a curious temple decorated with busts and grotesque animals, which 'doth very greatly exceed every artificial hermitage, grotto and cave that has yet been made or begun in this Kingdom'.[13] Langley's plans, submitted in June 1736, were for a bridge with nine arches, the centre arch of 120 feet, the remainder reducing by ten feet each towards the two shores.[14] It admitted 880 feet of waterway and was to be built at New Palace Yard rather than the Horseferry, where the river was nearly 300 feet narrower and the velocity of the water consequently greater. This was probably the most reasonable part of his scheme, the rest of his pamphlet being devoted to brushing aside the calculations of Labelye and Desaguliers and pouring scorn on the semi-circular arches of Hawksmoor's bridge. Even the great Palladio was attacked for his advocacy of semi-circular arches. Arches, according to Batty, should curve irregularly to obtain what he called a 'thurst' at right angles. The curves of his own arches, he claimed, 'have not any Part of a Circle, Ellipsis, Parabola or Hyperbola in them. Their abutments are within the Bases of their Piers and their strength is so great and so much superior to all other Arches of the same Dimensions, that weight which will break them down cannot have any Effect on these'.[15] 'Strange curves indeed', said John James, 'that have not any part of a Circle in them!' John James, the architect of St George's, Hanover Square, and surveyor of Westminster Abbey, at a later stage produced a design of his own for Westminster Bridge which was warmly recommended by Ripley, mainly because it was a timber bridge and therefore unlikely to be adopted. However at this point he contented himself with publishing a pamphlet reviewing the designs offered by Hawksmoor, Price and Langley. The first two have already been noticed (page 52). Batty Langley's design he demolished so effectively that it was scarcely even considered by the commissioners. Ripley reported that he had made so few arrangements to keep the timber dry that 'in a short time I apprehend the whole will come to ruin'.[16] It is interesting to note, however, that Batty Langley did recommend the use of 'concave parallelopipedons' instead of the more usual coffer dams for building the piers. 'Concave parallelopipedons' turned out to be timber vessels, strongly built with double joists and planks, within which 'workmen may work, altho' 15 or 20 Feet below the Surface of the Water, with the same Safety, Ease and Expedition as on dry Land'—in fact the caissons which were actually used by Labelye.[17]

In July a curious incident occurred in Westminster Hall, causing a good deal of alarm and confusion among the throngs of people attending the courts. A brown paper parcel was discovered on fire under the counsellor's bench in the corner of the court of Chancery. It was immediately kicked down the steps to the landing where it flared up and exploded, filling the hall with smoke and creating such consternation that, in the words of Lord Hervey, 'everybody concluding it was a plot to blow up the Hall, the judges started from the benches, the lawyers were all running over one another's backs to make their escape, some losing part of their gowns, others their periwigs, in the scuffle, and such an uproar it occasioned that nobody thought his own life was safe or knew how it came to be in danger'.[18] When peace was restored it was found that copies of the Bridge, Gin, Smuggling, Mortmain and Sinking Fund Acts had been parcelled up, filled with gunpowder and set on fire. During the confusion a number of papers had been scattered about the courts, announcing that five 'finished books or libels, called Acts of Parliament' had been publicly burnt at the Royal Exchange, Westminster Hall and Margaret's Hill, Southwark. The five Acts were enumerated, the Westminster Bridge Act being described as 'An Act to prevent Carriages and Passengers coming over London Bridge, to the great detriment of the Trade and Commerce of the City of London and the Borough of Southwark'—a colourful account, for there was nothing in the Act to hinder traffic over London Bridge.

The Lord Chancellor laughed heartily and cried *Clameur de Haro,* but the Lord Chief Justice, Lord Hardwicke, took a serious view and said the affair was more important than the ordinary business of the courts and in fact approached High Treason. The Middlesex Grand Jury, on being shown the papers, unanimously brought in a presentment of 'a wicked,

# THE STATE LOTTERY.

The Name of a LOTTRY the Nation bewitches, | The Footman resolves, if he meets no Disaster,
And City and Country run Mad after Riches: | To mount his gilt Chariot, and vie with his Master:
My Lord, who already has Thousands a Year, | The Cook-Wench determines, by one lucky Hit,
Thinks to double his Income by vent'ring it here: | To free her fair Hands from the Pot-hooks and Spit:
The Country Squire dips his Houses and Grounds | The Chamber-maid Struts in her Ladies Cast Gown,
For Tickets to gain him the Ten Thousand Pounds: | And hopes to be dub'd the Top Toast of the Town:
The rosie-jowl'd Doctor his Rectorie leaves, | But Fortune alas! will have small Share of Thanks,
In quest of a Prize, to procure him Lawn-Sleeves | When all their high Wishes are bury'd in Blanks:
The Tradesman, whom Duns for their Mony importune | For tho' they for Benefits eagerly watch'd,
Here, hazards his All, for th'Advance of his Fortune: | They reckon'd their Chickens before they were hatch'd.

Engraven By B. Roberts & Sold by him at his Shop in Ball Alley Lombard Street

1739

Price 4ᵈ

1739

8. The State Lottery, 1739 – the third Bridge Lottery at Stationers' Hall. (*British Museum Political and Personal Satire 2435. Photograph by courtesy of the British Museum*)

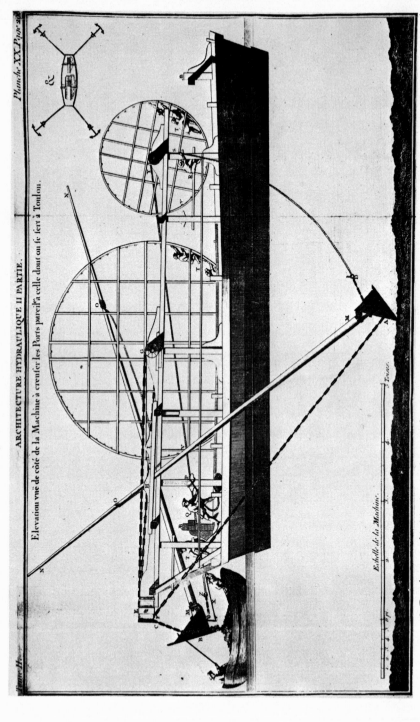

9. Dredger used in Toulon harbour. *(Plate XX in Belidor's Architecture Hydraulique (1753). Photograph by courtesy of the Royal Institute of British Architects)*

false, infamous and scandalous libel, highly reflecting on His Majesty and the Legislative power of this Kingdom, and tending very much to alienate, poyson and disturb the minds of His Majesty's subjects'. A proclamation offering £200 reward for discovering anyone concerned in the outrage was publicly read at the Royal Exchange and shortly afterwards three printers were taken into custody on suspicion. On Sunday 15 August two of the King's Messengers arrested the Reverend Robert Nixon, a non-juring clergyman of Hatton Garden, and took him before the Duke of Newcastle at the Secretary's Office in Whitehall where he was charged on oath as the author of the libel and committed to Newgate. A Mr Spittle, prisoner in the rules of the King's Bench, was seized at his lodgings in Southwark, and also a printer's journeyman. Nixon was admitted to bail before the Lord Chief Justice, tried and found guilty on 7 December, to receive sentence on the first day of the next term. He was sentenced on 10 February 1737 to make a tour of Westminster Hall and into the four courts with a paper on his forehead declaring his offence; to pay a fine of 200 marks; to suffer five years' imprisonment; and to find two sureties in £250 each and himself bound in £500 for his good behaviour for life. He performed the first part of the sentence immediately and then was remanded back to Newgate where we lose sight of him. He seems to have been quite crazy, acting on his own responsibility. Even when on bail awaiting sentence he had to be re-arrested for publishing a seditious libel on the birth of the Old Pretender. His connection with Westminster Bridge is remote but the affair created some stir at the time and the trial was thought worth publishing as *The Tryal of Robert Nixon . . . for a High Crime and Misdemeanour in publishing a Seditious Libel on Five Acts of Parliament*.[19] It was symptomatic in a period of unease typified by riots, anti-Irish disturbances and an underlying dread of Jacobinism.[20]

During the summer and autumn of 1736 there took place a number of meetings which were extremely ill-attended, due partly to the commissioners spending holidays in the country away from the noise and squalor of London, and partly to a vicious epidemic of riots in Westminster, Lambeth, Southwark and Spitalfields, caused by the 'Gin Act' taking effect from Michaelmas Day 1736.[21] At these meetings, 'there not being a sufficient number to act pursuant to the powers given by the Act', the clerk could do no more than determine a date for the next meeting. At last, on 16 February 1737, they achieved a quorum with the faithful chairman Lord Sundon, the Bishop of Rochester, Colonel Bladen, and Messrs Winnington, Earle, Crosse, Oglethorpe, Kent, Corbett, Hucks, Plumtre, Eliot, Whitworth, Walker and Treby. The main business was soon decided. It was a petition to allow the commissioners to extend the scope of the bridge lottery, which had raised only £43,116 of the £625,000 aimed at by the Act.

The petition, which was laid before the House of Commons on 3 March 1737, said that one of the main causes of the lottery's failure to raise the money was the want of encouragement by way of premium. If more time were allowed and more prizes offered, the commissioners believed a bridge lottery might prove successful, especially if it were reinforced with a toll similar to the Fulham Bridge toll. On the other hand, as far back as August of the previous year, Sir Thomas Robinson, MP for Morpeth and a commissioner in the Excise, had written to Lord Carlisle blaming it on the City.

> So 'tis thought there's an end of that scheme for raising money; 'tis owing to the Citizens being unanimous in not putting into it, and influencing all they can others being concerned in the affair.[22]

By the end of the year it was clear that the lottery was doomed unless it were immediately reorganised. A writer in *The Old Whig* (23 December 1736) reverted to the idea of annuities at 8 per cent with a toll assigned for payment which would obviate the need for a lottery altogether. A committee of the commissioners examined several witnesses to try to discover the reasons for the failure, and one of them, a broker called Richard Shergold, attributed it mainly to the moneyed people not thinking it worth while to engage. He suggested that a premium of 2 per cent should be offered to subscribers of fifty tickets or more.[23] This and others of Shergold's ideas were incorporated in the next lottery, which was more successful.

The new lottery was explained in the second Bridge Act, which authorised the commissioners to raise £700,000 by means of 70,000 £10 tickets.[24] Various ingenious incentives were incorporated in the Act, no doubt suggested by Shergold. Adventurers were to be allowed £15 on every fifty tickets bought; sums of money subscribed to the first lottery could be carried over to the new one; bearers of tickets were allowed a better chance for £7 10s od. The prizes were reduced in number—7,000 as against 30,000—made up of one of £10,000, two of £5,000, three of £3,000, six of £2,000, eighteen of £1,000, thirty of £500, ninety of £100, 200 of £50, and a larger number of £20 prizes, besides £500 to the owner of the first ticket to be drawn and £1,000 to the owner of the last ticket. The whole amounted to £227,500, which added to the £472,500 of the blank or unfortunate tickets, made £700,000.

Should the lottery prove unsuccessful the Act authorised the Crown to incorporate the trustees and commissioners, giving them powers to raise money for the bridge by charging pontage ranging from two shillings for coaches, chariots, berlins, chaises, chairs or calashes drawn by six or more horses, down to sixpence for every drove of calves, hogs, sheep or lambs. Foot passengers were to be charged a penny on Sundays and a halfpenny on weekdays. The toll money was to be vested in the trustees who were empowered to borrow money on credit of the Act at a rate of not more than 5 per cent.

The draw began at the Stationers' Hall on 14 November 1737, the first ticket winning £500 for a journeyman saddler called Bass, formerly a porter at the Young Man's Coffee House. It continued until 19 December, the £10,000 prize being drawn on 1 December to a ticket held by Mrs Robert Mire of St Mary Axe who had given a half share to Mrs Lafarell, widow of Colonel Lafarell, and the other half to two maiden ladies of Throckmorton Street, Miss Elizabeth and Miss Magdalen Torin.[25]

During the year the lottery generated a fierce controversy in the *Whitehall Evening Post* and the *Daily Post* over the value of tickets and the chances of a prize, one side declaring the lottery to be 'a Scheme for Gaming at 65% Loss' run by the touts of Exchange Alley, their opponents working out the odds at 14 per cent and demonstrating, from a scrutiny of the lists at the Bank of England, that the wildest gamblers were the City merchants and tradesmen. Others admired the guile of the citizens of Westminster in taking advantage of the folly of the country to build their bridge for them. Most serious-minded people, aware of the dishonesty of the draws at Stationers' Hall and the Guildhall and concerned at the degrading effects of gambling on a large proportion of their fellow countrymen, believed that some sort of legislation should be brought in, if only to control the plague of private lotteries attached like parasites to the bridge lottery.[26]

## NOTES TO CHAPTER 4

1. Act 9 George II, c 29 (1736).
2. The MPs included the Members for London and Middlesex, Surrey, Kent, Sussex and the Cinque Ports.
3. Labelye *Description* p 66.
4. The Act named the thirty managers and recited the oath 'that I will faithfully execute the Trust reposed in me ...'
5. For this and other information on lotteries see C. L'Estrange Ewen *Lotteries and Sweepstakes* (1932).

6. Acts 10 & 11 William III, c 17 (1699).
7. For subsequent Bridge Acts see Bibliography page 297.
8. The list was headed by the Duke of Grafton, the Duke of Newcastle, the Earl of Pembroke, Lord Burlington, the bishops of Chichester and Rochester, Onslow, Walpole, Pelham, Wager, Wade, Lord Charles Cavendish, Bladen, Crosse, Walker and Corbett, most of them faithful attendants at meetings over the next few years.
9. Bridge Minutes 22 June 1736; *Daily Post* 23 June 1736; *Political State* 52 p 13 (July 1736). The models have unfortunately disappeared.
10. Contracts (4) 22 June 1738.
11. Letter from Charles Knowles to Hermand 5 July 1736 (MS in RIBA library).
12. For Blackerby see *Gentleman's Magazine* I 209, II 874 977, III 551, VII 316 451, XII 219 275 331; *Calendar of Treasury Books and Papers* (1735–8) pp 128, 270, 445, 619; Bridge Minutes passim to 1742; St Margaret's Vestry Minutes 1725–42.
13. J. P. Malcolm *Londinium Redivivum* IV (1807) p 172.
14. Batty Langley *A Design for a Bridge at New Palace Yard, Westminster* (5 June 1736).
15. Ibid p 19.
16. Ripley's report in Bridge Minutes 24 August 1737.
17. James *Short Review* and Langley op cit p 26.
18. Lord Hervey *Memoirs* II p 311.
19. *Daily Post, London Evening Post, Read's Weekly Journal,* 15–17 July 1736, *London Tatler* 25 December 1736, *Political State* 52 pp 112–19, *Gentleman's Magazine* VI (1736) pp 398–400, 421, 485–6 and VII (1737) p 121. The incident is also described in Coxe *Memoirs . . . of Sir Robert Walpole* III p 348 and Philip Yorke *Earl of Hardwicke* (1913) I pp 137–40.
20. George Rudé *Hanoverian London 1714–1808* (1971) ch 10.
21. On 18 August Lord Sundon and six others appeared at the Bridge Office; on 31 August Lord Sundon and two others, Eversfield and Corbett; on 15 September Sir William Clayton (a cousin of Sundon), Joliffe and Eversfield; on 29 September only Sir William Joliffe turned up, and on 13 September only Sundon himself; on 27 October nobody (except of course the clerk, who was invariably there), and on 10 November only the Bishop of Rochester and John Plumtre; on 24 November nobody, and on 2 February 1737 Lord Sundon and four others.
22. Sir Thomas Robinson to Lord Carlisle, Albemarle Street, 11 August 1736 (*Hist. MSS Comm. Carlisle 6, 15th Report* p 174).
23. CJ 16 March 1736/7, vol 22 pp 807–8.
24. Act 10 George II, c 16 (1737) sections 2–30; *Gentleman's Magazine* VII (1737) p 314.
25. *Daily Post* 15 November and 2 December 1737.
26. *Political State* 53 and 54 (1737) throughout; *Common Sense or the Englishman's Journal* 25 June 1737 quoted in *Gentleman's Magazine* VII (1737) pp 367–9.

# 5
# Ripley versus Labelye, 1737-8

Meanwhile, on 16 March 1737, Lord Sundon had reported to the House of Commons on the Bridge Commission's work over the past year. Sir Joseph Ayloffe had reported on the lottery and Richard Shergold had made his suggestions for the second lottery. The Comptroller of His Majesty's Works, Thomas Ripley, had been questioned at length on the cost of building a bridge at New Palace Yard compared with Fulham Bridge, on the difference in cost and endurance of stone and timber bridges, and on what the cost might be of driving a free passage to New Palace Yard and a road across the Southwark marshes.

Ripley's costings are interesting. He produced two estimates for a stone bridge from New Palace Yard, the first, built on a solid bed of wooden piles eight to ten feet deep, costing £80,000 and taking three summers or possibly only two; the second, built on solid stone piers, costing £100,000 for the foundations alone. A stone bridge from the Horseferry would cost £75,000. Timber bridges were more difficult to predict because of the depth of water at Low Water Mark, but he estimated possibly £26,000 at New Palace Yard and £20,000 at the Horseferry. He was asked how long a timber bridge would last and what the cost of repairs might be over a period of fifty years, and answered that it would be equal to its being built twice over in the same time, provided it were as well made as the Fulham Bridge. He also produced a compromise, half timber and half stone, by building stone walls two feet thick on two rows of dovetailed piles, laying timber from pier to pier and waterproofing with layers of lead, which would protect it for probably fifty years with only small repairs. This would cost, including the purchase of land for the approaches, at New Palace Yard about £38,000, and at the Horseferry about £30,000.[1]

Sundon then presented the Bill to the House which discussed it in committee on 5 May, negativing a proposal to buy out the Fulham Bridge proprietors. The Bill was agreed by the Lords and received the Royal Assent on 21 June 1737.[2]

Up to this point only the site of the bridge and its cost had been discussed. The first serious deliberation on its design occurred at a meeting of the commissioners on 6 July 1737 with Lord Sundon presiding and twenty-eight members present. To assist them they called in Ripley himself, the masons Jelfe and Tufnell, the carpenter John Barnard, the Swiss engineer Charles Labelye, the two blacksmiths Joseph Pattison and Thomas Wagg, the plumber George Devall, and the lighterman Samuel Price, all of whom were well-known in the building trade. They considered Ripley's plans for a stone bridge of fifteen arches with a thirty-foot roadway and recesses on each side 'for people to rest in or get out of danger'. Ripley, who had succeeded Vanburgh as Comptroller of the Works in 1726, had already had experience of designing—if not actually building—a bridge, having submitted plans for Fulham Bridge in 1726 and 1728 (see pages 39–42). He was an undistinguished architect, a fact that can be verified by a glance at the Admiralty, but Pope's cruel couplet of 1737,

> Who builds a bridge that never drove a pile?
> (Should Ripley venture, all the world would smile),

is witty but unduly harsh and doubtless relates more to Ripley's dependence on the patronage of Walpole than to his professional ability, which, however, Pope would have been perfectly competent to judge.[3] As Horace Walpole, who quotes the couplet, shrewdly observes, 'the truth is politics and partiality concurred to help on these censures. Ripley was employed by the Minister and had not the countenance of Lord Burlington, the patron of Pope'.[4] Certainly Jelfe and Tufnell and the other builders were prepared to work to his design for the sum of £120,000, of which they proposed to deposit a £40,000 security in the commissioners' hands until two years after the centres had been struck.

Ripley had first put forward his theories to the commissioners at a meeting in the Jerusalem Chamber in June, though a month later the estimate had risen from £75,000 to £90,000. He suggested

> a solid bed of piles dovetailed together and drove 8 or 10ft deep or as deep as the nature of the soil would permit, and the heads to be 18 inches below Low Water Mark, from thence wholly stone and 30ft wide in the clear with the recesses on each pier . . . a convenient Bridge for Man, Horse and all sorts of Carriages and might be completely finished in three summers if not two.[5]

However his estimate of £35,000 for a timber bridge at the Horseferry appealed to the commissioners a great deal more. They passed two resolutions immediately to that effect and published an advertisement in the *London Gazette* and other papers saying that anyone wishing to contract for the bridge should deliver their plans and proposals by 29 July.[6] This allowed a bare three weeks but as London was humming with talk of the new bridge it is to be assumed that most candidates already had their schemes well under way. Certainly at the commissioners' meeting on 21 July several plans appeared and a committee was appointed to examine them and report. This committee, known as the Works Committee of which five was to be a quorum, met for the first time on 2 August 1737 in the Jerusalem Chamber under the chairmanship of Joseph Danvers, MP for Totnes.[7] Plans and models were submitted by Samuel Perry, James King, William Westley and his brother John Westley, John James, John Henslow and John Lock, Charles Labelye (here spelt Laballey), John Wyatt, Batty Langley, William Bartlett (a timber bridge of seven arches) and Charles Marquand (laid aside as impractical). The Works Committee did not meet again until the following June.[8]

A fortnight later Danvers reported to the commissioners that his committee, with the help of Ripley, had examined several plans, models and designs, but that there had seemed to be so many uncertainties and difficulties that they had resolved to suggest that the commissioners should undertake the bridge themselves under proper direction. Ripley then reported on the various entries, none of which, in his opinion, were good enough. Some were too insubstantial to resist the damp and ice; none made provision for keeping the joists dry; Henslow and Lock proposed making the main piles of iron 'which I think very improper'; and Wyatt and Bartlett's design allowed only 700 feet of waterway instead of the 760 feet required by the Act. Only James's timber bridge met with Ripley's approval. It was perhaps not surprising that his recommendation was for the two middle piers, the middle arch and the two abutments of his own design to be erected as an experiment. He produced a model of the two piers on 31 August and expounded his ideas to the board. Meanwhile Labelye was brought in to show the commissioners his device for laying the foundations of stone piers below the surface of the river-bed. He made several demonstrations to illustrate his idea and then introduced to the commissioners a ballast man, Robert Smith, who could level the bottom for them, having had previous experience with the York Buildings Water Company. Labelye and Smith then withdrew and the board resolved that in case they should decide to lay the foundations of a stone pier, Labelye would be a proper person to be employed and James King a suitable carpenter to assist him. It looked as though Ripley's bridge would win the day.

Up to this point the meetings had been fairly sparsely attended, usually between twelve and twenty commissioners turning up. During the winter of 1737–8, Wager, Ayloffe and one or two others decided it was high time someone determined the exact position the bridge was to take. Letters were circulated and advertisements published and on 8 February a meeting took place with fifty-one members present and Admiral Wager in the chair. A memorial from the inhabitants of Westminster was read out repeating their petition for the bridge to be built as near the centre of Westminster as possible. The chairman was ordered to ask Parliament for a decision the next week. This took the form of a petition from the Bridge Commissioners, telling the Commons first that the original plan was to build the bridge from the Horseferry but that the general feeling in London favoured New Palace Yard or the Woolstaple; secondly that they were precluded by the terms of the Act which allowed New Palace Yard only and 'the Petitioners desire the House will please to determine the situation of the said Bridge'. Petitions were then presented from Westminster praying that the bridge should not be built from the Horseferry, and from the parishes of St John the Evangelist, St George's Hanover Square, St James's and St Margaret's, praying that it *should* be built from the Horseferry which had always been thought by 'the most skilful Mathematicians, Navigators, Surveyors and Artists' to be the only fit place for such a bridge. The House then resolved itself into committee and decided 'that the place for building a Bridge most advantageous to the Trade of the City and Liberty of *Westminster* and most commodious to the inhabitants thereof, is from the Woolstaple or thereabouts . . .'[9] Sundon, Winnington, Wager, Lord Baltimore and William Pulteney were then directed to prepare a Bill, which was read for the first time on 17 March and for the second time on 29 March, ordered to be engrossed on 12 April with an amendment enlarging the commissioners' powers, agreed by the Lords and received the Royal Assent on 20 May 1738.[10]

The third Bridge Act was entitled 'An Act for Building a Bridge cross the River Thames from the *Woolstaple* or thereabouts, in the Parish of Saint *Margaret* in the City of *Westminster* to the opposite Shore in the County of Surrey'. It was short and to the point and apart from authorising the commissioners to build at the Woolstaple instead of New Palace Yard—a matter of a few yards only—the one clause of fresh interest was that allowing the lottery earnings to be lent out at the Receipt of the

Exchequer on public security for the benefit of the bridge. The commissioners had achieved one decision—the site. The actual material out of which the bridge was to be built, whether of wood or stone, was still left unresolved and the problem was causing Londoners a good deal of concern. Ripley's compromise of a timber bridge on stone piers had appealed to the commissioners for reasons of economy, but except for King and Barnard (the carpenters who hoped to win the contract for the superstructure), most people regarded a wooden bridge in the heart of the capital as unworthy of a great nation. Desaguliers had deplored the possibility in the House of Commons from the technical angle, but from the point of view of both prestige and aesthetics a timber bridge at Westminster half way through the eighteenth century was laughable. 'A bridge at Westminster or somewhere betwixt the present Bridge and Lambeth, is now the affair at hand', said a writer in *The Prompter* in 1737,

> and without controversy, will prove *egregio publico*—Yet tis but the other day that there were certainly found, treading upon English Earth, creatures strange enough to petition against a new Bridge . . . So absurd is it to be against a second Bridge that we ought rather to aim at a third. There are three at Paris: and I think as many at Rome (cities much less extensive than London), nearer one another than three would be here . . . I'll only add that I hear this Bridge, we expect, is after all to be only a WOODEN one. I am resolved not to believe that till I see them driving the Piles. Oh Shame! a wooden Bridge! a vile paltry *Pons Sublicius* (no better perhaps than that at Fulham) over the Thames in the sight of London!

*The Daily Post* fervently hoped that there was no truth in the report, only too general, that 'it will end in a wooden Bridge and built by the Tolls being assigned in perpetuity—the one not for the Honour of the Nation and the other the greatest hardship on the People who will in such case be liable to a perpetual duty'.[11] Labelye himself, in his description of the early stages of the bridge, said the public in general were disgusted at the thought of a timber bridge in the 'Metropolis of the British Empire'. He submitted a design of his own for a stone bridge of thirteen semi-circular arches and two small abutment arches, publishing an engraving by Fourdrinier the following year[12] (Plate 5).

However, at a meeting on 10 May 1738, the commissioners decided that the bridge should be of timber on stone piers and it was not until a year and a half later that public opinion over-ruled them. At this meeting Labelye again produced his model of a machine for laying the foundations of the piers, estimating the cost of thirteen machines at £14,172. These machines were the celebrated caissons invented by Labelye and described later (pages 93–94). Messrs Jelfe and Tufnell then delivered their estimate for building thirteen solid Portland stone piers to Labelye's design at £36,972, and the commissioners resolved that

> Mr Charles Labelye be employed as Engineer in laying the foundations of the piers, That he direct in what manner and scantling each pier shall be built, and That he take care to direct the foundation of the stone piers be built of such sufficient strength and dimensions as if a stone superstructure were to be placed thereon, and such a one as he had designed for a stone Bridge agreeable to his own Plan.

This was a definite statement in the direction of a stone bridge but they qualified it later on in the same meeting by a resolution that the superstructure be built of timber according to James King's design.[13]

King's bridge is worth examining briefly (Plate 12). If public opinion had not been so firmly opposed to a timber bridge it might well have been built instead of Labelye's stone bridge, the design of which had been sent in at the same time. Labelye described it as a 'curious and extremely well-contrived *wooden Superstructure,* of the Invention and Design of the late Mr James King. . . The model of this superstructure was then shown and explained to a great many Persons of the first Rank in the Kingdom, and to many others, and generally approved of'.[14] A large engraving of it, dated 22 June 1738 and dedicated to the commissioners, shows a thirteen-arched bridge rather flimsy in style and slightly like Walton Bridge, built about ten years later by William Etheridge for Samuel Dicker.[15] It was to have cost £28,000 and to be of 'as good Oak Timber as can be procured, 988 feet in length, 44 feet in breadth from out to out, and 39 feet in height in the middle from the underside of the kerb which is to lie immediately on the stone piers to the upper side of the rail, and 19 feet or thereabouts in height at the two ends'.[16]

John Barnard, a carpenter of Kensington, asked to be associated with King and the partnership continued until 1741 when Barnard was dismissed. James King remained the works carpenter until his death in 1744 when he was succeeded by his foreman, William Etheridge. Ripley's criticism of this bridge (or possibly an earlier model) was that it might 'at first prove very strong but there is in it so great a number of joints which cannot be kept dry that I think it would very soon come to decay'.[17] Ripley was of course an interested party, having sent in a design himself, and his criticism should be taken with circumspection. But judging from the engraving and from detailed drawings by one of Andrews Jelfe's foremen, Thomas Gayfere,[18] Ripley may well have been right in this case. Certainly Walton Bridge lasted for only a few years, the timber arches being replaced with brick in 1780.

An even more curious design was laid before the board at this time by James Turner, a merchant of Queen Street, Moorfields, who explained to them his method of building a bridge of two enormous arches of 375 feet each, a 50 foot pier in the middle, and 200 foot abutments. He was ordered to lay his design before Sir Jacob Ackworth (of Fulham Bridge) and any other mathematician he should think proper, and attend the board with a certificate of his abilities. Nothing further was heard of Turner.

The appointment of Charles Labelye as engineer to Westminster Bridge immediately raises the question of why a comparatively young and untried foreigner should have been given the task of undertaking such an important work. Labelye was born at Vevey, the son of François Dangeau Labelye, a French Protestant who had fled to Switzerland after the Revocation.[19] He had come to England some time between 1720 and 1725, and according to Vertue he worked as a barber. When he was examined as an expert witness by Wager's committee in 1736 he said that he had lived in England for sixteen years; and in the Preface to his *Short Account of the Methods made use of in Laying the Foundations of the Piers of Westminster-Bridge,* published in 1739, he modestly hoped that his readers would overlook his poor English, 'when they know that I was neither born nor bred in England but in Swisserland; and never heard a word of English spoken till I was near Twenty Years of Age'.[20] In 1725 he joined a French Masonic Lodge in London, the Solomon's Temple Lodge in Hemmings Row, his sponsor being the distinguished natural philosopher, J. T. Desaguliers, also a Huguenot refugee. He spent the years 1727 to 1728 in Spain where he founded a Masonic Lodge, coming back to England in November 1728 to be thanked for his services to Freemasonry in Madrid. He was also thanked at a meeting of a Lodge at the Bear Inn, Bath, on 18 May 1733,

> for the many good offices, useful instructions and unnumbered favours the Lodge has received from their worthy Brother Charles de Labely, through his zealous endeavours to promote Masonry, they unanimously desir'd the R. W. Master to return him their Hearty Thanks in Form.[21]

In 1734 he supplied maps and surveys of the river to the 'public-spirited gentlemen' who were laying their plans for putting the Bill before Parliament; in the following year he was employed surveying the coast near Sandwich; his tidal calculations were incorporated in Hawksmoor's scheme for Westminster Bridge (see page 51); and in 1737 he was included among the five surveyors of the river, submitting one of the many plans for a timber bridge.

Beyond these sparse facts very little is known of his career in England before the board's approval of 'Mr Labelley's design' and their resolution, dated 7 September 1737, 'that he will be a proper Person to be employ'd in case the Commissioners proceed to the laying the Foundations of a Stone Pier'. It is probable that he learned his profession of marine civil engineering in harbour works and fen drainage. He refers in the *Short Account* to his having watched different methods of laying the foundations of moles, dykes, ramparts and other works in the sea and fen country, and there are undated papers in the Bodleian Library referring to Ramsgate Harbour and the Eddystone Lighthouse. In the 1740s, when he had become well-established as a civil engineer, he published reports and surveys on draining the Fens and building harbours at Great Yarmouth and Sunderland.[22] But the key to his appointment over the head of the Comptroller of the King's Works probably lies at Sandwich, where in 1735-6 a new harbour was projected to try to reduce the toll of shipwrecks in The Downs.[23] The two Members of Parliament for the Port and Town of Sandwich, the notorious Sir George Oxenden and the secretary of the Admiralty, Josiah Burchett, commissioned Labelye to carry out a survey of The Downs and the coast between the North and South Forelands 'where ships bound westward are frequently detained by contrary winds a long time in The Downs, which is an open Road, where they are exposed to great Danger of Enemies in time of War and of being shipwrecked in stormy weather'.[24] In 1738 he published his survey, which included a canal and basin near Sandown Castle with sluices connecting to the river Stour. But in spite of various petitions to Parliament nothing materialised until the Ramsgate Harbour Act was passed in 1749, when Labelye's services were no longer called for—probably because of the subsidence of one of the piers of Westminster Bridge the year before.[25]

However the experience he had gained was enough to convince the commissioners that Labelye rather than Ripley was their man. There must have been a good deal of in-fighting before their object was achieved. Ripley was, after all, a protegé of Walpole, and through his patronage had become Comptroller of the Works from 1726 and keeper of the king's private roads, gates and bridges from 1737, among other offices. He was an architect in the prime of life with to his credit the Custom House, the Admiralty and two large private houses in Norfolk—Houghton and

Wolterton. But bridge building was very different from house building, demanding specialised techniques in tidal waters, in which Labelye had had some experience, and Ripley none apart from the brief flirtation with Fulham Bridge in 1728. Nothing in England existed at this time to compare with the famous *Corps des ingénieurs des ponts et chaussées,* founded in France in 1720 and developed by Trudaine into the *École des ponts et chaussées.* In fact, as Samuel Smiles wrote some hundred years later,

> our first lessons in mechanical and civil engineering were principally obtained from Dutchmen who supplied us with our first windmills, watermills and pumping engines . . . The art of bridge-building had sunk so low in England about the middle of the eighteenth century that we were under the necessity of employing the Swiss engineer Labelye to build Westminster Bridge. In short we depended for our engineering, even more than we did for our pictures and music, upon foreigners. At a time when Holland had completed its magnificent system of water communications, and when France, Germany and even Russia had opened up important lines of inland navigation, England had not cut a single canal, whilst our roads were about the worst in Europe.[26]

The prevalent idea that Labelye's appointment gave offence to English architects appears to be exaggerated. There was no lack of competition for the new bridge but most of the pamphlets alleged to reflect professional hostility were published in 1736 (see page 52), a year before the commission's resolution approving his design and two years before the resolution of 10 May 1738 appointing him engineer in charge of the foundations. Batty Langley's 'Swiss Impostor' pamphlet, showing Labelye hanging from a gallows beneath one of the arches (Plate 37), did not appear until 1748 when a defect had developed in the bridge foundations. But there is no doubt that jealousy did exist and Labelye went to some trouble in the *Short Account* to refute the critics, answering their objections one by one and demonstrating that his method of building the piers was cheaper and more efficient than any other,

> to the great Mortification of many evil-minded Persons, especially some disappointed *Projectors* and *Artificers*, who without knowing what was really intended to be done, or being capable of putting it into execution, roundly asserted everywhere that 'this Method of Building was entirely impracticable, or at least would prove so expensive that the Charge of laying the Foundation of one single Pier would amount to more than the whole Expence of the Superstructure'.

He made no attempt to defend the economics of the bridge, leaving that side of the work to the comptroller and surveyor of the Bridge Works, Richard Graham, and simply pointing to the piers themselves, standing solidly across the river, as proof of his competence. 'As to the many false Reports and Insinuations that have been maliciously spread abroad, and all that can proceed from Hearts fill'd with *Envy, Hatred and Malice and all Uncharitableness,* I leave the Piers of Westminster Bridge to give them the strongest Reproofs and the most solid answers.'[27]

But Labelye still remains a shadowy figure of whom no personal description is known and of whom, in an age when the art of portrait painting flourished, not even an engraving exists. Batty Langley's 'insolvent, ignorant, arrogating Swiss' is clearly the product of a soured and slightly lunatic mind. Horace Walpole dismisses 'the ingenious monsieur Labelye' in two lines. For a clear picture of the man we must turn to the six volumes of the Bridge Commissioners' Minutes and the two volumes of Works Committee Minutes. Labelye's name occurs, without a blemish, on nearly every page for the space of twelve years, ending with the award of £2,000 from the commissioners 'for his great fidelity and extraordinary labour and attendances, skill and diligence'. Sir John Summerson probably correctly imagines him belonging to that class of ingenious craftsmen and watchmakers flourishing in Switzerland, a professional type which England had not yet learnt to produce but which was shortly to develop into men such as Gwynn, Mylne, Smeaton, Telford and the Rennies.[28]

## NOTES TO CHAPTER 5

1. CJ 16 March 1736/7, vol 22 p 808.
2. Act 10 George II, c 16 (1737), 'An Act for explaining and amending the Act...'
3. Alexander Pope *Imitations of Horace* (1737) Book II Epistle I 185–6. See Colvin *Dictionary* entry for Ripley.
4. Walpole *Anecdotes of Painting* III p 106.
5. Bridge Minutes 29 June and 6 July 1737.
6. The voting at this meeting at first appeared to be equal at thirteen against thirteen and was determined by the chairman's casting vote; but eventually it transpired that Ayloffe had been counted in by mistake and the real voting was fourteen to twelve (Labelye *Short Account*) p 21.
7. The other members were Lord Sundon, Admiral Wager, Lord Charles Cavendish, Lord Gage, and Messrs Kent, Brereton, Crosse, Walker, Corbett, Whitworth, Revel and Papillon.

8. Works Minutes 2 August 1737. For Ripley's criticisms of the various entries, see ibid 24 August 1737. The plans and models have not survived.
9. CJ 16 February 1737/8, vol 23 p 38.
10. Act 11 George II, c 25 (1738).
11. *The Prompter* 24 February 1736/7, *The Daily Post* 6 February 1736/7, *Political State* September 1737 p 263. James Thomson (of *The Seasons*) published an indignant poem 'On the Report of a Wooden Bridge' in the *Gentleman's Magazine* August 1737 p 511.
12. 'An Explanation of the Plans, Elevations and Sections belonging to a Design of a Stone Bridge adapted to the Piers which are to support Westminster Bridge', Labelye *Short Account* pp 65–82 and *Description* pp 5–6.
13. Bridge Minutes 10 May 1738.
14. Labelye *Short Account* p 9 and *Description* p 7.
15. For Walton Bridge see the painting by Canaletto at Dulwich, Constable *Canaletto* 442.
16. Bridge Minutes 3 and 10 May 1738.
17. Ibid 24 August 1737.
18. Gayfere's meticulous watercolour drawings are in an *Account Book* presented to the Institution of Civil Engineers by George Rennie's widow.
19. Biographies of Labelye are by R. B. Prosser in the *Dictionary of National Biography* and by H. M. Colvin *Dictionary of Architects* (revised edition 1978).
20. CJ 16 February 1736/7, vol 22 p 569; Labelye *Short Account* p vi.
21. *Ars Quatuor Coronatorum* XL (1927) pp 37 and 244.
22. *The Results of a View of the Great Level of the Fens* (1745); *The Result of a View and Survey of Yarmouth Harbour taken in the Year 1747* (Norwich 1775); *Report relating to the Improvement of the River Wear and Port of Sunderland* (1748).
23. Survey and Map of Sandwich Harbour 1735/6 (Sandwich Records, Accounts 1735–36); W. Beys *History of Sandwich* (1792) pp 762–3; Dorothy Gardiner *Historic Haven* (1954) p 208.
24. CJ 2 March 1736/7, vol 22 p 766.
25. John Smeaton *Historical Report on Ramsgate Harbour* (1791) pp 2 and 10.
26. Samuel Smiles *Lives of the Engineers* (1861) I vii.
27. Labelye *Short Account* pp 37–8.
28. Sir John Summerson *Georgian London* (1970) pp 96–8.

# 6

# Building Begins 1738-9

Labelye's appointment as engineer dates from 10 May 1738. At the same time the commissioners appointed Richard Graham to be surveyor and comptroller of the Works at £300 a year, and Thomas Lediard to be surveyor and agent at £200 a year. They both worked faithfully for Westminster Bridge for many years, Lediard until he died in 1743 and was succeeded by his son (also Thomas Lediard), and Graham until he died in 1749 and was succeeded by Obadiah Wylde. They relieved Labelye of the burden of measuring the works, managing the accounts and contracts, and negotiating the sale of lands and houses needed for the bridge approaches, leaving him free to choose building materials and to direct the works. His own salary was fixed at £100 a year and a subsistence of '10 shillings per diem without respite for his constant attendance', to take effect from 10 May 1738. [1]

The comissioners then turned their attention to settling contracts. deciding on Messrs Jelfe and Tufnell, masons of Westminster, for building the two middle piers, and on the carpenters Messrs King and Barnard for the timberwork. The masonry agreement for the two middle piers was signed on 22 June 1738 between Ayloffe and Blackerby for the commissioners, and Andrews Jelfe and Samuel Tufnell for the masons who were called on for the sum of £1,000,

> to find and provide a sufficient quantity of Portland Stone of such scantling and sizes as shall from time to time be directed and approved by Charles Labelye, Gent. the present Engineer . . . setting up, erecting, building, compleating and finishing the said two largest stone piers . . . each to be in length sixty-one feet, in breadth seventeen feet and in height twenty-nine feet . . . in good and workmanlike manner compleated . . . within the space of twelve calendar months. [2]

The partnership of Jelfe and Tufnell is interesting. [3] Andrews Jelfe came from an Essex family and became an experienced mason as clerk of the works at Newmarket and clerk itinerant in the Office of Works, and by

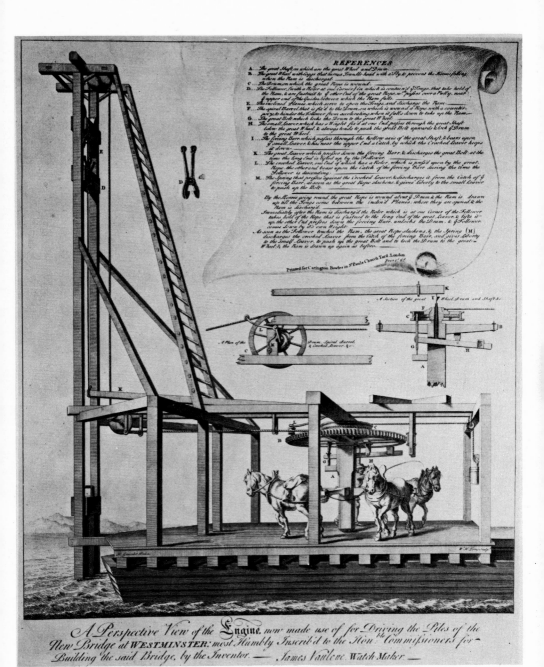

10. Vauloué's pile-driving engine. (*Engraving by Gravelot by courtesy of the Trustees of Sir John Soane's Museum. Photograph by R. B. Fleming*)

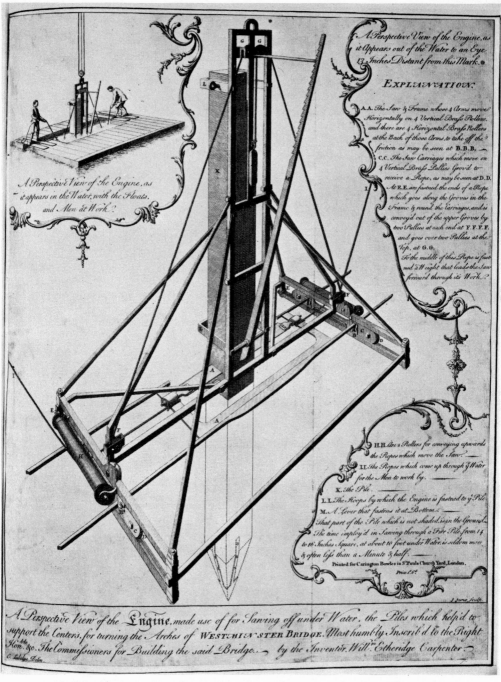

11. Etheridge's machine for cutting piles under water. (*Engraving from a drawing by Labelye by courtesy of the Trustees of Sir John Soane's Museum. Photograph by R. B. Fleming*)

winning the masonry contracts for several of the 'Fifty New Churches'. His most valuable experience relating to Westminster Bridge was gained in Scotland where, in the service of the Board of Ordnance, he built or repaired a number of garrisons, forts, castles and probably military bridges. His wharf was in New Palace Yard. His daughter Elizabeth married Captain Griffen Ransom, whose timber yard at Stangate had 'with Kindness and Civility' been freely offered when the foundations were being laid. He was elected one of the gentlemen of St Margaret's Vestry on 8 September 1749.

Samuel Tufnell came of a well-established Westminster family, following his father and grandfather as master mason to Westminster Abbey.[4] He was a vestryman of St Margaret's from 1729 with a house in Old Palace Yard, and a pillar of the militia (eventually becoming a Colonel), and in 1751 he was judged to be among the thirty 'discreet and substantial inhabitants of the Parish of St Margaret's Westminster and assigned to be Commissioners of the Court of Requests for the City and Liberty of Westminster'.[5]

Meanwhile, on the site of the bridge itself, a marker pile had been driven in the middle of the river to mark the position of one of the piers for the centre arch, Labelye having reported to the commissioners that he had bored the river bed where the middle arch on the Westminster side was to stand and had found hard gravel at about five feet below the mud.[6] Then, early in July, he and the masons went down to the Isle of Portland to choose stone, selecting the Grove Quarry as a likely source and setting the hands (mostly from London) to cutting and working into scantlings enough stone for the two middle piers. A wharf to receive the stone was found in Ransom and Smith's timberyard on the Surrey side near Stangate, and preparations were immediately started for driving piles and building frames for the caissons.

Labelye's task was considerably lightened by a special machine for driving piles invented in about 1737 by a 'very ingenious Watch-Maker of my Aquaintance', James Vauloué. Previously pile-driving plant had been of the most primitive order, consisting of crude contraptions motivated by man-power and slow and laborious in the extreme. Perronet had invented an engine driven by river or tidal current at about this time but Vauloué's machine certainly seems to have been one of the first.[7] Desaguliers, in a lecture on hydraulic machinery, describes Vauloué's engine in detail, illustrated with an engraving by J. Mynde (Plate 10). 'In works where we are confin'd to Time by Tides, as in making and mending Bridges', he says, 'there must be great care to husband the Power so well as to lose no Time. The late Mr Vauloué, Watch-Maker, contriv'd the best of that kind that perhaps was ever seen, and about 5 times more expeditious than the best I ever heard of.'[8]

Vauloué's machine (the Engine, as it was called) was driven by three horses walking round a windlass and, by means of a system of gears and ratchets, raising a heavy weight or ram between ten and twenty feet into the air. The ram was then released by a trigger and descended on the pile, which had been temporarily strengthened on the top with an iron hoop. Three horses could deliver about five strokes of the ram in two minutes, and a thirty-four foot firwood pile, thirteen or fourteen inches square and pointed with an iron spike, could be driven about fourteen feet into the river bed in an hour or so, saving a great deal of time and expense. A working model, formerly in George III's scientific collection and used for the instruction of his children, can be seen in the Science Museum.[9]

The three horses were provided by Robert Halliwell, who agreed to stable them near the Woolstaple, feed them on oats and beans, furnish harness made of the best material, and provide a man to drive and look after them. In return the commissioners agreed to ferry them to the bridge site and to pay Halliwell £120 a year plus £15 12s od for harness. The contracts were renewed every three months from September 1738 until August 1745 and again in July and October 1749.[10]

This engine was a great help. It was begun in July by James King's carpenters who were given five shillings overtime pay on 29 August for staying out till eight o'clock to test the engine, and a guinea on 14 September—'to 20 of Mr King's men employ'd on the Engine being the first day it was used'.[11] This was for the first of the guard piles or fenders, used to protect the works from damage by barges, and by the end of October, with the help of Vauloué's engine, they had driven enough guard piles to protect the works for the two middle piers—thirty-four long ones and twenty-two short ones for the first pier and twenty-six long ones for the second, with floating booms rising and falling with the tide. After it had been working successfully for six months or so, Vauloué pointed out that his engine was in constant use 'and humbly hopes that the Honble Board will think him entitled to a suitable encouragement'. Labelye and Graham were asked for their views and both said without hesitation that it was greatly preferable to any other engine they knew. Vauloué was then awarded a gratuity of £150[12] The engine can be seen at work in various paintings of the bridge under construction (Plate 22).

Meanwhile the carpenters were at work in the shallow water near the Stangate, constructing the caisson within which the first pier was to be built. Labelye's caisson was another device designed to simplify the work. Hitherto the usual method of building piers had been by means of coffer-dams or *batardeaux,* which involved driving hundreds or even thousands of piles to form a cavity within which foundations could be laid. This was a time-honoured engineering technique used not only for bridges but also for fortifications, moles, embankments, jetties, sluices and other water-works, sometimes built singly (*batardeaux simples*), sometimes double with clay rammed between the rows of piles (*batardeaux doubles*). It had been employed from Roman times to the great age of French bridge building under Louis XIV and was used again by Rennie for Waterloo Bridge and in more technically sophisticated forms by civil engineers ever since. Labelye's caissons were a refinement of the English chests, which were timber boxes filled with stonework and sunk in a rather haphazard way to the bottom. Caissons were enormous flat-bottomed boats, eighty feet long and thirty feet wide, which could be sunk accurately, raised again by pumping, and resunk as often as necessary. Labelye was extremely proud of them, though he did not claim to have been their inventer. 'The Method I have made use of is not absolutely new', he wrote in the *Short Account.* 'Something like it is often practised in the Construction of *Moles, Dikes, Ramparts* and other Works erected in the Sea or in Lakes, or in deep Fens or Morasses, and I have assisted in making use of it, where all other methods have prov'd ineffectual.'[13]

The most primitive caissons were baskets of reeds filled with stones. These developed into wooden chests hooped with iron. The great French civil engineer Gauthey, relating the history of Mansard's Pont Royal, built after the floods of February 1684 had washed away Mazarin's timber bridge, describes the use of a kind of *caisse* or *crêche* large enough to contain several courses of masonry for an entire pier. The *caisse* was then sunk over a previously dredged foundation and used thereafter as a coffer-dam. It would seem therefore that the builder of the Pont Royal, Friar François Romain, a Dominican monk who had been working earlier on the bridge at Maestricht, was the first engineer to use large caissons. This is con-firmed by the Pont Royal accounts, from which it can be shown that Romain had been employed as director of works practically from the beginning and not, as Gauthey says, after the emergency of one of the piers subsiding.[14]

Labelye's caissons differed from Romain's and the English chests in the greater accuracy to which they enabled the masons to site the piers, and in their immense size. Chests had usually been about sixteen feet square. Caissons were eighty feet long and contained 150 loads of fir timber 'of more tonnage and capacity than a Man of War of 40 Guns'. To make this scale even more impressive to the imagination, the ground plan of one of the middle piers was chalked out on the floor of the Painted Chamber in the Palace of Westminster, 'and there was hardly room to go along the Walls without treading upon the outward Lines though that room is full 66 feet by 25; so that the Inside of one of the Caissons in which the Piers were built . . . is very near the same as the Painted Chamber'.[15] Romain's *caisses* were as large as Labelye's but could not be pumped out and refloated. Batty Langley claimed that he had first thought of the idea and that his 'concave parallelopipedons', published in his *Design for the Bridge at New Palace Yard, Westminster* in 1736, were in fact exactly the same and 'that he had been villainously robbed of the Honour and Profits thereof, justly his Due'.[16] Labelye did not deign to reply, merely repeating that the caisson method was nothing new and simply a development of existing methods of hydraulic engineering. Larousse, however, credits him with the invention—'ce système de fondation parait avoir été employé pour la première fois à la construction des piles du Pont de Westminster'—and most civil engineering text-books concur, from Charles James's *Military Dictionary* (1810) onwards.

While the caisson was being built, Robert Smith, ballast man of Lambeth, was excavating and levelling the foundation pit to the same shape as the caisson and five feet wider all round. He employed two gangs of men and contracted to dredge the foundations for each pier in forty-two days. His contracts were drawn up for individual piers, the first being signed on 12 October 1738, the last on 1 June 1743, the masonry for the fourteenth and last pier being completed in February 1744.[17] Thereafter he continued in the commissioners' service, providing ballast for the new Surrey road.

Smith probably used a barge fitted with long pivoting poles worked by two wheels, one large and one small, driven by men or boys climbing round inside. This was the harsh method used for dredging Toulon Harbour in 1750 (Plate 9) and a variation of it was no doubt used for Westminster Bridge. Accurate levelling was determined by means of measured gauges in the form of eighteen-foot poles attached to stone anchors and painted red and white in feet and inches from the bottom of the stone.[18] By this method Labelye reckoned to level the bottom to within one inch. To prevent silting up the pit was protected by boards let into grooved piles a few feet from the edge. But silting was not the only hazard. Labelye complained to the Works Committee that some watermen had run their barges foul of his boats moored alongside Vauloué's engine. The clerk was ordered to advertise in the evening papers and at the landing places for a fortnight, quoting the Act that 'if any person shall wilfully and maliciously blow up, pull down or destroy the said Bridge . . . [he] shall be adjudged guilty of Felony and shall suffer Death as a Felon without Benefit of Clergy'. This was only moderately successful because on New Year's Eve, during the night, 'some ill-designing persons had pulled out the plug of a Barge belonging to Joshua Smith, ballast man employ'd in the Works'. The barge had sunk and a reward of ten pounds was offered to be paid on conviction of the offender. [19]

A week later the foundation of the first pier was finished, 'to the great satisfaction of Mr Labelye, Engineer, Mr Graham, Surveyor, and several Gentlemen of distinction'. On 15 January the first caisson was launched and moored in position between the guard piles, the men being rewarded with five guineas. On Monday 29 January 1739,

> in the Afternoon, the first Stone, which weigh'd upwards of a Ton Weight, for building the Main Arch of the new Bridge, was laid by the Right Hon. the Earl of Pembroke, Groom of the Stole to His Majesty, with great Formality, Guns firing, Flags displaying, &c. [20]

Labelye rather touchingly tells us that the Earl of Pembroke 'coming accidentally to view the Works, condescended to set the first Stone; which was the middle Stone of the Foundation of the first Pier. Soon after which his Lordship retired, leaving ample Marks of his wonted Generosity among the Workmen'. [21]

Pembroke had long been interested in Westminster Bridge (Plate 16). His father, the eighth earl, was a celebrated virtuoso and connoisseur, acquiring the Arundel Marbles and forming most of the collection of pictures at Wilton House. Locke had dedicated to him the *Essay concerning Human Understanding*. Henry Herbert, the ninth earl, inherited his talent and became an amateur architect rivalling Lord Burlington. [22] The Palladian bridge at Wilton (1736–7) was, according to Vertue, 'the design of the Present Earl of Pembroke & built by his direction'; and Horace Walpole declared it was 'more beautiful than anything of Lord Burlington or Kent'. [23] Queen Caroline described the earl as 'the best creature in the world and meant very well . . . but he is as odd as his father was, not so tractable, and full as mad'. This did not preclude his playing a prominent part throughout the building of Westminster Bridge, from the formation of the commission in 1736, of which he was frequently chairman, to the ceremonial laying of the first and last stones, and indeed until the day of his death in 1750 when he had attended a meeting of the Works Committee that very morning. Horace Walpole declared that 'it was more than taste, it was passion for the utility and honour of his country that engaged his lordship to promote and assiduously overlook the construction of West-minster-bridge by the ingenious monsieur Labelye, a man that deserves more notice than this slight encomium can bestow'.[24] Pembroke's name never actually appeared in the early stages of the negotiations for West-minster Bridge but from later evidence there seems to be every likelihood that he was directly concerned with it, and probably among the 'distin-guished Gentlemen' and the 'enlightened Noblemen' responsible for pushing the Bill through Parliament. The bridge officers frequently brought their troubles to him, knowing that he was both influential and sympathetic to their cause. Labelye, in sending him a progress report in the summer of 1739, frankly says that 'no person is more Sensible than I am that it is owing entirely to your Lordship that a Bridge shall be erected at Westminster'. And Graham sent him another progress report in October ending with much the same words—'tis owing to your Lordship that there is a Bridge & that I am concerned in it. I thought it my Duty (as I know you have the success of it so much at Heart) to give your Lordship this information'.[25]

After Pembroke had laid the first stone and the guns and fireworks had ceased, Labelye reported to the board with his customary modesty that 'the Case for building the first Pier was moor'd in its proper place and that a great quantity of stone was weigh'd into her and some part already set'.[26] The work went ahead at speed. Shiploads of stone arrived punctually from Dorset, the apparatus for transferring it from the barges to the masons inside the caisson worked efficiently, the caisson itself was raised and lowered several times to ensure absolutely accuracy in the foundation, on 30 March the sides of the caisson were floated off (the men being supplied with nine shillings' worth of victuals to prevent them going ashore) and on 23 April 1739 the scaffolding was struck to reveal the first pier complete, 'without loss of either Life or Limb in any of the Workmen'.[27]

They had encountered difficulties, of course. In January and February two more barges had been deliberately sunk in order to foul the caisson and more warning notices had to be stuck up at the watermen's plying places. Cornwall, the waterman owning one of the barges, was awarded seventeen shillings damages; Robert Smith, the ballast man owning the other, petitioned the Board for damages 'done to my Barges and of Goods lost out of them since I began to make the Foundations of the first Pier . . . not thro any neglect but by spitefull Malicious persons'.[28] Rumours were going round that the foundations were badly laid and that the caisson had not been sunk level, and Labelye had to reassure the board that the pier was founded on hard gravel, that the caisson was level and that no serious damage had befallen the works. He concluded his report, in fact, by declaring that 'no Pier was ever better set'.

The piers took from three to six months to build according to weather conditions and the supply of stone. Contracts with the masons were signed for the two middle piers in June 1738, for four more in June 1739, for another six in January 1741, and for the last two in April 1742.[29] The timber shears and crab used for hoisting stone were usually removed on completion but if one can judge from paintings of the bridge by Scott, Jolli and others, the last pier, completed in February 1744, retained its timber work for some time afterwards (Plates 25–27). Usually at this time piers were built hollow with an outer shell of stone filled inside with brick and rubble. Labelye's piers were entirely solid, built of regular courses of Portland stone, 'all the joints filled with a Cement made of Lime and Dutch Tarris which sets and hardens in water . . .'[30] (Plate 7).

Various administrative problems were naturally to be expected and Sir Joseph Ayloffe's accounts give a vivid picture of the day-to-day details coped with by the clerk from December 1738 to December 1739.[31] The accounts begin with payments for contingencies such as £9 for advertisements throughout the year and for sticking notices at the plying places after the barges were sunk, six guineas for Mr Poyntz, committee clerk to the House of Commons, ten guineas to the Attorney General and the Solicitor General for consultation on the clause enabling the king to sell any part of the Crown lands in Westminster that might be needed for the approaches, nineteen shillings to Robert Vaile for mending a drain, and so on. Ayloffe left a number of blanks to be paid for at the commissioner's discretion, for instance 'to drawing the Bill to enlarge the powers of the Commissioners to enable them to raise money by Lottery—very long and difficult I submit to the Com$^{rs}$'. And he added a memorandum on the difficulties of seeing the two Bills through both Houses.

> I had frequent attendance upon the Committees of the House of Commons . . . attending several evenings till 12 o'clock at night at Sir Charles Wager's when they met to settle the Bills, attending Lord Sundon at his own house, several times attending the Attorney and Solicitor General . . . and all other matters relating to the soliciting of the said Bills in which a good deal of time was spent and expenses incurred for which I do not presume to make any charge but submit the same to the consideration of the Commissioners.

Other items include £64 to Mrs Hannah Bristow for the entertainment of the jury deciding on Mr Cumberland's estate, 2s 6d for the waiters, and £11 to Samuel Richardson for printing the Bridge Bills.

These were minor problems and fairly easily solved. The main obstacle, and one which was to prove a stumbling-block for years to come, was the legal conveyance of land and buildings in order to clear a way for the bridge approaches. Details of estates, leases, juries' decisions, verdicts and valuations, can be found in the three relevant volumes in the Public Record Office.[32] Thomas Lediard's detailed reports were transcribed into the Bridge Minutes. Some twenty pages of Ayloffe's accounts refer to the conveyance of various estates: Mrs Penn, George Lucey, Thomas Whincap, the Parish of St Margaret, the Christ's Hospital estate in New Palace Yard, and so on. His expenses usually involved the customary legal charges for perusing title deeds, making fair copies, searching the wills at Doctors Commons, paying the Middlesex Register for searching for incumbrances, drawing bargains, engrossing leases, and paying for parchment and stamps. Occasionally Ayloffe was careful to point out that he had made no charge for certain services and he sometimes emphasised the obstacles he had to overcome:

> I attended Mr Stackhouse, one of the mortgagees, three times to prevail on him to join in conveyance and deliver me the title deeds . . . I attended Mr Brochett at Christ Hospital four times in relation to the Title and called or sent to him 6 or 7 times in regard to the execution . . . I attended Mr Pounteney twice at Kensington to inform myself as to the 3rd part of his estate which appeared to belong to his wife's sister.

Entries in the Vestry Minutes for April 1739 confirm that the parishioners of St Margaret's were only too anxious to help towards the construction of a bridge, which all agreed would be immeasurably to their advantage. Provided the conveyances were fair to the occupants and that no annuity holder should suffer unduly, they were prepared to meet Ayloffe and agree willingly to the commissioners' plans. A Vestry order of 10 April 1739 empowered the churchwardens to treat with the commissioners for the conveyance of parish houses in King Street and the Woolstaple. The transaction was complicated by some of the estate belonging to the Woolstaplers' Company. In order to clarify this part Ayloffe searched the Auditor of the Land Revenue's Office to find the company's title to the houses on the Westminster Market, and the records in the Tower of London and the Rolls Chapel for the company's Charter of Incorporation. He also spent several hours at Mr Heaton's house at the further end of Barnaby Street, Southwark, to make an extract of the company's charter from Charles II.

The St Margaret's part of the Woolstaple estate was happily resolved when

> Mr Churchwarden Haselar reported that he with his partner, John Lawton Esq., in pursuance of an Order made last Vestry (for treating and agreeing with the Commissioners) had mett Thomas Lediard Esq. Agent for the said Commissioners and were acquainted by him that the Commissioners were willing to give Twenty Six Years purchase for the said Ground Rents in the Woolstaple and would reserve to the Parish the said Annuities the better to answer the intention of the respective Donors. And this Vestry, having considered the said Report, Do unanimously agree to Accept Twenty Seven Years purchase for the said Ground Rents in the Woolstaple, and the said Thomas Lediard Esq. attending was acquainted therewith; and the Churchwardens are hereby desired and Impowered to agree with the Comm[rs] accordingly . . .

The money was used by the Vestry to pay workhouse repair bills and for immediate poor relief.[33] Ayloffe's quietus was approved and signed by seven of the commissioners (Pembroke, Cavendish, Brereton, Wade, Laroche, Plumtre and Wager) who allowed him £31 10s od for the expenses left blank, the total for the year 1739 amounting to £1,056 16s 11d.

Thomas Lediard had been appointed 'Surveyor of the Ways, Streets and Passages, and Agent for buying and selling such estates as the Com[rs]. shall find necessary to buy or dispose of, with a Salary of £200 p.a. commencing on 22nd June 1738'.[34] He was undoubtedly a remarkable man and the results justified the board's choice. He had been a great lover of travel, had served as a young man on the Duke of Marlborough's staff, had been attached to the Diplomatic Service, and had written a number of books.[35] His portrait, surrounded by an assortment of devices representing Geometry, History, Cosmography and the like, appears as frontispiece to his *Naval History of England,* and shows a thin-lipped portly gentleman enjoying the throes of inspiration by Nautica, the muse of naval history (Plate 17).

In the early stages of the negotiations for Westminster Bridge, Lediard lived in Smith Square and was a member of the 'Society of Gentlemen' which arranged the meetings in the Horn Tavern in 1734.[36] At that time he had combined with the magistrate, Thomas Cotton, who had already prepared a scheme for improving the layout of King Street and the approaches to New Palace Yard, Westminster Hall and the Abbey. Lediard,

Hawksmoor, 'a Gentleman of great Worth' and Cotton himself had then improved and enlarged the scheme; but Hawksmoor had died, Cotton's career as a magistrate and 'the private concerns of his Profession' had allowed him no time to spare, the 'Gentleman of great Worth' was overcome by a multiplicity of other public affairs, and so 'the labouring Oar as well as the Expense of taking Surveys, drawing Plans, making Calculations and procuring other necessary Intelligences' fell wholly on Lediard's shoulders. [37]

Certainly by the beginning of 1738 he was deeply involved in the bridge affairs and in February published over Cotton's name and his own *A Scheme . . . for Opening Convenient and Advantageous Ways and Passages to and from the Bridge,* printed on a single quarto sheet accompanied by a plan. [38] Briefly the *Scheme* which the authors humbly offered to the commissioners involved enlarging New Palace Yard by opening it to the Woolstaple, demolishing a large number of houses north of New Palace Yard between King Street and the river, and building a new street, ninety feet wide, from New Palace Yard towards Charing Cross as far as the Admiralty Office. Most of the houses were ruinous and Lediard calculated that, allowing sixteen years' purchase, the total cost would amount to £64,320. The streets were narrow, the houses dilapidated and 'unfit for Business or Pleasure and consequently inhabited chiefly by the meaner Sort of People'. The new development would change Westminster from a slum into a residential and commercial district 'inhabited not only by Persons of great Distinction, for the Convenience of being so near the Parliament House and Westminster Hall, but by rich and eminent Tradesmen of every sort'. The authors concluded their *Scheme* with a résumé of the ground rents that could be expected, and with a plan of the area from Westminster Hall to the Plantation Office and the Privy Garden.

Lediard sent a copy of the *Scheme* to one of the commissioners, who answered in a letter dated 25 February 1738, phrased in a critical and sometimes offensive tone of language. He taxed Lediard with not mentioning the earlier proposals, with taking the credit to Cotton and himself only and ignoring the others concerned who were 'more equal to it than yourself', with making no provision for traffic blocks near the Palace of Westminster, and with not providing a market place. He concluded by accusing Lediard of being party to a 'Sort of Tyranny, to oblige People to quit their Habitations and Possessions, whether they will or not'. The writer concealed his identity behind the initials N.N., which (not applying to any of the commissioners) presumably stood for No Name or some other anonymity. Perhaps he was the Gentleman of great Worth. [39]

This was a godsend to Lediard, who immediately took the opportunity to answer at some length and with considerable patience, only occasionally allowing a note of acerbity to escape and reveal his feelings for N.N.'s egregious criticisms. He published all three, the *Scheme*, N.N.'s *Letter* and his own *Answer*, together with a brief proposal to establish a perpetual fund for maintaining the bridge by means of rents drawn from a new Westminster market which, though touched on only lightly in the *Scheme*, was elaborated in the *Answer* and the accompanying plan (Plate 2). This pamphlet appeared in April 1738 and was addressed to the Members for Westminster, Lord Sundon and Admiral Wager. It was called *Some Observations on the SCHEME offered by Messrs Cotton and Lediard* and was a convenient way of drawing the attention of the commissioners to his interest in the bridge affairs. It was very probably the main cause of an appointment which was to test his abilities, stamina and tact for the next five years. In fact he died at work in 1743.

Lediard's duties were for the most part concerned with hard bargaining for property which had to be demolished in order to make way for the bridge approaches. The Bridge Minutes are packed with his reports recounting the briefest outlines of negotiations and interviews with a wide variety of property owners. His experience as a writer enabled him to compress into a few pages the results of haggling which had occupied him for weeks and sometimes months at all hours of the day and night. The reader, sifting through the many volumes of the Bridge Minutes, is always enlivened by Lediard's reports and indeed longs for more. If only he had had the time to write a full account of his negotiations he could have produced a social history of the greatest interest.

Some of the estates he resolved were those of Christ's Hospital ('after a great number of fruitless attempts to come to an agreement with the Govs of Christs Hospital for their estate in Westminster Market'), Mrs Elizabeth Ekins who owned a shop, five houses and a tenement on the north side of New Palace Yard, the Reverend Dennis Cumberland of Northampton who owned the White Horse alehouse in Cannon Row and three houses adjoining it on the north side of the Woolstaple, Carter's carpenter's shop, the Woolstaple almshouses whose four residents were willing to accept two shillings a week for lodgings until they could be properly housed, and innumerable private dwellings. Sometimes settlements were made easily, sometimes they dragged on for months and could be resolved only by empanelling a jury. Typical settlements were £325 to Mr Cumberland, £1,100 to Mrs Ekins, £525 to Thomas Whincap for two houses in King Street, £525 to the Governors of Christ's Hospital for a house in New

Palace Yard, £360 to Robert Pountney for a house in King Street, £105 to Hercules Taylor for a loan of the Woolstaples estate, and £351 to the Parish of St Margaret's for ground rent on three houses in the Long Woolstaple. For months Lediard treated with John Phillips, a poulterer, for his freehold property in Union Street and King Street, but found him 'so buoyed up by his neighbours Lewis, Cartwright and Potter, that he will not abate one penny of his first demand of £1,600'. Lediard had originally valued the property at £1,215 but because the houses were old and standing on very little ground had reduced it by £100. 'It is my humble opinion', he reported to the commissioners, 'that as his exorbitant demand is founded on the expectation of a Coronation, the prejudice which will accrue to him in his trade of a poulterer . . . I shall never agree with him without a jury.' As it turned out, however, they came to terms, finally agreeing on £1,260.[40]

Sometimes outgoing tenants took with them all sorts of landlords' fixtures such as locks, doors, windows, wainscotting and lead gutters, 'which if not taken notice of may encourage others to do the like . . . and occasion a considerable loss'. Empty houses were robbed so often that Lediard was ordered to sell all the lead and portable material as soon as possible after settlement and to make inventories beforehand. He found it a great strain and at the end of one prolonged negotiation he was unable to report its outcome in person, being 'unhappily seized this morning with a fit of the Stone which renders me incapable of attending my Duty'.[41]

In fact Lediard began to find the work more and more of an effort and in 1740 he presented a long memorial to the board, pointing out that he was fifty-five, and had served as agent and surveyor for two years, and the work had increased to such an extent that he had been compelled to employ an assistant to carry the surveying instruments. He also employed his son Thomas, who had drawn some 200 plans for the use of the juries. He himself had been collecting rents on behalf of the commissioners and on account of the number and poverty of most of the tenants, this occasioned him 'a great deal of pains and labour'. Furthermore he had been obliged almost continually

> after a laborious day's attendance upon business, to spend the whole evening in writing and calculating, by which means as he can undertake no other business, your Memorialist has, during the two years he has served in these several employs, been a sufferer and expended a large sum out of his private fortune, over and above the salary allowed him.

Lediard's work and his petition were appreciated by the commissioners, who immediately awarded him an extra £100 a year and a bonus of £200.[42] Even so his labours overtaxed his strength and he died at work three years later.

## NOTES TO CHAPTER 6

1. Bridge Minutes 9 June 1738.
2. Bridge Contracts (4) 22 June 1738.
3. Colvin *Dictionary* entries for Jelfe and Tufnell.
4. E. B. Tufnell *The Family of Tufnell* (1924); F. W. Steer *Samuel Tufnell of Langleys* (1960) has apparently no connection.
5. St Margaret's Vestry Minutes 1 May 1751 p 274.
6. Bridge Minutes 9 June 1738; *Gentleman's Magazine* VIII (1738) p 322.
7. Perronet *Description des Projets* (1788).
8. J. T. Desaguliers *A Course of Experimental Philosophy* (1745) pp 417–8.
9. The engine is described in detail in Labelye *Short Account* pp 21–3 and in Desaguliers op cit p 417.
10. See Appendix pages 285–89.
11. Accounts 19 May 1738 to November 1739 in Bridge Minutes I.
12. Bridge Minutes 8 June 1739 and Bridge Accounts p 19.
13. Labelye *Short Account* p 39. The caisson method is first mentioned in *Political State* September 1737 p 264.
14. E. M. Gauthey *Traité de la Construction des Ponts* (ed Navier 1809) I pp 69–70, and F. de Dartein *Etudes sur les Ponts en Pierres* (1907) II pp 77 and 84.
15. Labelye *Description* pp 86–7.
16. Batty Langley *Survey of Westminster Bridge as 'tis now Sinking into Ruins* (1748).
17. See Appendix pages 285–9.
18. Labelye *Short Account* p 26; Belidor *Architecture Hydraulique* (1753) II pp 156ff.
19. Works Minutes 23 August 1738 and 3 January 1739.
20. *Read's Weekly Journal* 3 February 1738/9.
21. Bridge Accounts 15 January 1738/9; Labelye *Short Account* p 29.
22. Colvin *Dictionary* entry for Herbert; Earl of Pembroke *Catalogue of Paintings and Drawings ... at Wilton House* (1968).
23. Horace Walpole's letter to Horace Mann, 10 January 1750.
24. Horace Walpole *Anecdotes of Painting* IV p 108.
25. Wilton Papers, letters from Labelye to Pembroke 26 July 1739, and from Richard Graham to Pembroke 13 October 1739.
26. Bridge Minutes 31 January 1738/9.
27. Bridge Minutes 26 April 1739; Labelye *Short Account* p 36.
28. Bridge Minutes 31 January 1738/9 and 12 December 1739.
29. See Appendix pages 285–9.
30. Labelye *Description* pp 19–20.
31. Quietus of Sir Joseph Ayloffe, Bodleian MS. Gough London 1.
32. PRO Works 6/60–2. A short list of properties purchased by the commis-

sioners between 1739 and 1750 (to the value of £91,817) is in Bridge Accounts pp 100–6.

33. St Margaret's Vestry Minutes 21 April 1739 p 18, and 12 June 1741 p 72.

34. Bridge Minutes 22 February 1738/9.

35. Among Lediard's books were a biography of Marlborough, the *Naval History of England,* a continuation of Rapin's *History of England* covering the reigns of William and Mary and Queen Anne, and an English Grammar he had published in Germany while secretary to 'His Majesty's envoy extraordinary in Hamburg'.

36. In 1743 he was living in Old Palace Yard paying £30 12s 6d in Poor Relief.

37. Lediard *Observations* p 11. Cotton was in trouble at this time for receiving money from informers on illegal gin-sellers (Middlesex Sessions Book No.951, April 1738).

38. For Lediard's *Scheme*, the commissioners' answer and Lediard's reply, see Bibliography 299–300.

39. Lediard *Observations* pp 7–9.

40. Bridge Minutes 28 January and 29 February 1739/40.

41. Bridge Minutes 23 March, 26 April, 8 June, 27 June 1739, and 7 May, 25 June, 3 September 1740.

42. Ibid 16 April 1740.

# 7
# The Great Frost, 1739–40

The infinitely complicated problems of property conveyance had been foreseen by the commissioners who, earlier in the year, had petitioned Parliament for a Bill to enlarge their powers and to grant them more funds, 'not only to finish the said Bridge but also to make and render Commodious several Ways, Streets and Passages to and from the intended Bridge, and at the same time to make several convenient Passages leading to the Courts of Justice and to both Houses of Parliament.'[1] Lord Sundon reported from the committee appointed to examine this petition and told the House that Thomas Lediard, who had been inquiring into the cost of buying up property with the aim of opening streets between the foot of the bridge and Whitehall, had been faced with objections from many landowners claiming that the commissioners had insufficient legal powers to compel them to sell. Lediard had in fact mentioned in one of his reports eight almshouses in the Lower Woolstaple and part of 'the Crown Waste in New Palace Yard which if possessed by the Commissioners may be of great use to bring two or three proprietors, who have actually encroach'd upon it and are hitherto extravagant in their demands, to a more reasonable way of thinking'.[2]

Sundon then presented the Bridge Accounts from which it could be seen that from the £98,000 raised by the lotteries, £20,375 14s od had been paid out in prizes, £5,853 to the Managers, £3,000 to the Bank of England, and £5,168 to several contracting workmen; £60,000 remained at the commissioners' disposal, of which £28,000 had already been set aside in contracts. Labelye had estimated the cost of the bridge to be £90,000, which could be broken down into £6,000 for the two largest piers, £37,000 for the thirteen other piers, £5,000 for the two abutments, £14,000 for apparatus for building the piers, and £28,000 for the timber superstructure. The committee said that more money would be needed and that more powers should be granted to the commissioners in order to remove all doubts about the purchase of property on either side of the approaches. Sundon presented the Bill on 10 April, the Chancellor of the

Exchequer informed the Commons that the king had said 'the House may do therein as they shall think fit', the Bill was passed on 22 May, agreed to by the Lords on 7 June and received the Royal Assent on 14 June 1739.[3]

The fourth Bridge Act was entitled 'An Act to enlarge the Powers of the Commissioners for building a Bridge cross the River Thames . . . and to enable them by a Lottery to raise the Money for the several Purposes therein mentioned; and to enlarge the Time for exchanging Tickets unclaimed in the last Lottery; and to make Provisions for Tickets in the said Lottery lost, burnt or otherwise destroyed'.[4] It was a help to the commissioners as far as it went, enabling them to negotiate with landowners and to empanel juries. It also authorised a third lottery to raise £325,000 in £5 tickets. There were to be 16,310 prizes, the first being two premiums of £10,000, the first and last tickets drawn to be £500 and £5,000 respectively, and the smallest prizes to be £10. Fifteen per cent of the £325,000 was to be deducted for the bridge; and the usual penalties were threatened for selling chances of tickets for less than the whole time of drawing and for selling interest in tickets not in the vendor's possession.

The draw began at Stationers' Hall on 10 December and ended on 25 January. The first ticket to be drawn, Number 23,969, won the prize of £500 for James Favey, partner with John Elli, 'an eminent Dealer in Irish Linen of Lawrence Lane, Cheapside'. Two £3,000 prizes were drawn by Miss Buckworth, a niece of Sir John Buckworth, and by a country carrier who had bought the ticket that morning, two hours before it came up. A £5,000 prize was won by two brothers, John and James Porter, merchants of Tokenhouse Yard. One of the £10,000 prizes was won by William O'Brien, formerly a Spanish merchant from Rotterdam, who had the grace to donate £20 of it to the sufferers in the great frost that was then at its height. The other £10,000 prize fell to some ladies in the families of Mason and Simpson, Jamaica merchants of Tower Hill. The last ticket was drawn on 25 January, winning £5,000 for Joseph Arnold, an attorney of Gainsborough, Lincolnshire, whose relations in London had bought it for him a few days earlier.[5]

By this time gambling, particularly private lotteries and raffles, had become a public menace, sapping the energy and industry of the country to such an extent that yet another Act had to be passed to try to control it.[6] Henry Fielding commented in *The Champion*,

> Nothing being more reasonable than that the Vices of private Persons should contribute as much as possible to the advantage of the Public; the Game of *Passage* is to be suppressed this year, as that of Hazard was the Last;

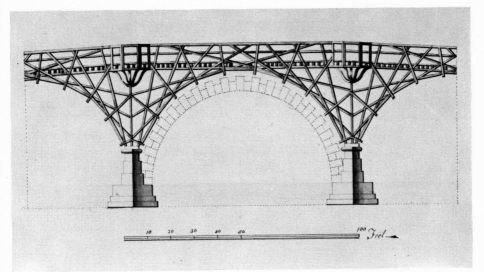

12. James King's wooden superstructure, 1738. (*Drawing by Thomas Gayfere in the Institution of Civil Engineers. Photograph by R. B. Fleming*)

13. James King's timber centring. (*Drawing by Thomas Gayfere in the Institution of Civil Engineers. Photograph by R. B. Fleming*)

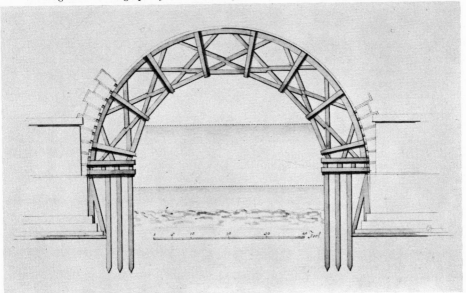

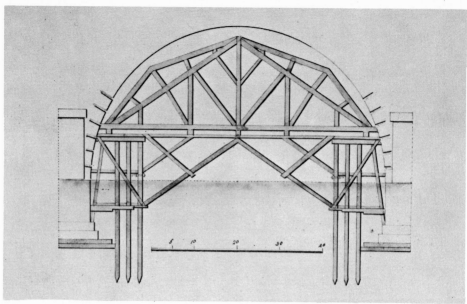

14. Labelye's timber centring erected for the damaged arch in 1750. (*Drawing by Thomas Gayfere in the Institution of Civil Engineers. Photograph by R. B. Fleming*)

15. View through one of the Etheridge's timber centres, 1746–7. (*Engraving by Parr after Canaletto published in June 1747. Photograph by courtesy of the British Museum*)

the Consequence of which, tis hoped will be, that *Annual State Lotteries* will from henceforth glean up all the Money which used to be confounded at Dice; and in due time furnish Posterity with a Monument at Westminster that may be called THE BRIDGE OF FOOLS through all Generations.[7]

During the spring and summer of 1739 the work had gone ahead smoothly, the first middle pier being finished in April, the second in August, both according to contract.[8] Ten new commissioners were elected to replace deceased members, but from June onwards board meetings became sparsely attended, the officers being liable to be held up for want of authority.[9] In May and June, Jelfe and Tufnell had contracted for the two abutments and the half piers and the four piers on either side of the middle piers already built.[10] But difficulties were not far off. Strong easterly winds blew almost continuously, building up the high tides and reducing the time available for work in the caissons. The price of timber rocketed. An agreement had been signed with a timber merchant, Richard Stevens of Shad Thames opposite the Tower, for Nerva and Riga timber, the price in 1738 to be £2 a load for Riga from thirty to forty feet long, and £1 18s od for Nerva or Mardo timber. By midsummer 1739 a sub-committee (Cavendish, Hucks and Laroche) noted that 'the price of Fir and Deal advanced about 5% and Baulks and Yew Firs about 7% and that Mr King had advanced his price as far as possible in pro-portion'. By December they reported to the board that timber had doubled in price since the 1738 valuation. The increase led to an outbreak of timber stealing. Notices stuck up at the watermen's plying places offered a reward of five shillings for every pile brought to the commissioners' wharf at Stangate, and five guineas for the discovery and conviction of any timber thief.[11]

During the winter of 1739-40 an exceptionally hard frost seized the country and great damage and suffering were caused by ice and snow. The frost began on Boxing Day growing more severe even than the memorable winters of 1683-4 and 1715-16 when the Thames froze solid and oxen were roasted whole on the ice. 'Many who had lived for years at Hudson's Bay', said a writer in the *Gentleman's Magazine*, 'declared they never felt it colder in those parts. The Thames floated with Rocks and Shoals of Ice . . . Multitudes walked over it and some were lost by their Rashness.' The frost was not of course confined to the Thames. Coaches drove over the ice on Windermere. A printing press was set up on the Ouse at York. The Newcastle citizens played football on the Tyne. The river Liffey was frozen as far up as Ringsend. In Germany and Poland wolves and bears from the frozen forests roamed the villages scavenging for food. And in

Russia, on the shores of the Neva, the empress built a palace twenty feet high entirely out of blocks of ice and guarded by six ice cannons which could actually fire six-pound ice cannon balls.[12]

In the early stages the frost was treated light-heartedly and the novelty of walking about on the river, playing skittles and buying drinks and coffee from the Queen's Head Punch Booth, the Frost Fare Coffee House, the Friezland Coffee House and other stalls, offset the discomfort of the cold. One could buy crude colour prints of the scene at the King's Head and Theodore's printing booths; and an edition of Moseley's famous engraving of the law courts in Westminster Hall, 'The First Day of Term', was actually printed in a press on the ice above London Bridge. Teams of boys dragged rafts and boats from one side to the other offering rides to any hardy soul who would venture. Mr Cunningham of Fulham wagered £20 he could gallop his horse from Fulham to Hammersmith and back in an hour. He won his bet in forty-five minutes. One of the Chelsea water pipes burst and spouted up under a willow tree, freezing amongst the branches and forming 'a perfect Arbour of Crystal to the infinite admiration of all who beheld it but being soon converted into a Gin Shop, every one who purchased a Dram took a Fragment of this extraordinary Fabrick away with them so that already hardly the Ruins of it remain'. A sheep was roasted whole on the ice near Pepper Alley Stairs, 'whence great numbers of the Unthinking Giddy flock'd to see the novelty but some of them paid dear enough for their Folly, for by crowding in an extraordinary manner several were pushed in.'[13] There was a well-worn path across the ice from Whitehall Stairs to the Southwark shore with a side track to the two middle piers of Westminster Bridge where sightseers could mount a ladder and stand perilously on top of the newly built masonry (Plate 19).

But very soon the novelty wore off and real hardship began. The Chelsea watermain was not the only one to burst. Most of the London water pipes were lengths of hollowed tree trunks which easily split when choked with ice. Water flooded out and immediately froze, cracked the pavements and made walking perilous. Watchmen and lamplighters, often old and infirm, were reluctant to venture out and the streets were plunged into even deeper darkness than usual, with the inevitable increase in robbery and murder. The son of Herman Van der Myn, painter to the Princess of Orange, was drowned while skating on St Marylebone reservoir. Coal became practically unobtainable, dozens of colliers being laid up below London Bridge. The watermen's trade was at a standstill, their boats gripped in the ice or stove in and sunk, and by the end of the month more than thirty boats had been lost between Tower Stairs and Woolwich. The Kent and Essex shores were dotted with flocks of mallard, easterling (smew), widgeon and coot, either perished with cold or starved to death. On the first Sunday in January three boys ventured out in a boat off Rotherhithe and 'were drove away by the great Flakes of Ice and perish'd thro the severity of the Frost'. The poor suffered terribly and though there was no system of national relief, many individuals and private organisations contributed to help the worst sufferers. The king gave £4,000 to be distributed among the poor housekeepers of London and Westminster. The Duke of Somerset and the Duke of Newcastle gave food and huge sums of money. Sir Robert Walpole gave seventy guineas for the poor of Westminster and Chelsea. Most parishes made collections to help the worst cases and the curate of St Margaret's collected £76.[14] Even William O'Brien, who had won £10,000 in the bridge lottery, sent a £20 bank bill to the vicar of West Ham relieving 300 familes with beef, bread, coal and money. Large groups of carpenters, bricklayers and other workmen, with the tools of their trades draped in mourning, walked through the streets imploring assistance for their families. And a party of watermen and fishermen actually carried a peter-boat festooned with black drape round and round St Paul's beating a dead march on the drums. The newspapers were full of such stories and, with a large proportion of the population living on the verge of starvation anyway, it is not difficult to imagine the privations endured by a great number of people.[15]

Towards the end of January the inhabitants of the houses on London Bridge opened their windows one morning to find the sterlings cluttered up with a motley collection of booths, shops and huts brought down from Westminster—a booth with trinkets, a shed with the dregs of a dram of Old Gold, a skittle frame and pins, and a stall lettered 'the Noble Art & Mystery of Printing'. This was a hopeful sign but the thaw did not seriously begin until well into February. On Sunday 17 February, 'a channel was opened and the Watermen in various parts ply'd and carried Passengers in common, 'tho Sunday, which was winked at by the Rulers in Pity to them who have lately suffer'd so great Hardships'. Shortly afterwards some oyster boats and three Kentish corn hoys appeared below Tower Steps and by the end of February ships were coming up the river again in glut.[16]

The great frost was over. The damage it had inflicted on the works of Westminster Bridge was extensive but not more than might have been expected. Before the winter the two middle piers had been finished, a third was well under way, the Westminster abutment was level with Low

Water Mark, and the foundations of the Surrey abutment had been begun. The two piers, protruding solitary and forbidding above the ice, can be seen in Griffier's 'View of the Frozen Thames' in the Guildhall and in several Frost Fair engravings (Plates 18–19). From Christmas until 18 February all work was totally stopped and by the time a channel had been cleared and it was possible to examine the situation, they found the ice had carried off all the piles then in use (about 140) and the sides of one of the caissons had been shattered.[17] At a meeting of the commissioners on 23 February Graham reported that the water bailiff had 'siezed and marked for his own use several of the Piles belonging to the Works of the Board, a barge loaded with stone, the cassoon for building the Piers, and other materials belonging to the Works that had been drove away by the late bad weather'. Graham was told to get it all back forthwith and 'demand of him his reasons for seizing them and putting his mark thereon'. It turned out however that the bailiff had simply rescued them on behalf of the commissioners and Graham was ordered to 'return the thanks of the Board to the Water Bailiff for the great care he took of the materials belonging to the Works'.[18]

The stonework of the piers and abutment was comparatively unharmed and the damage to the piles and caisson was repaired by 19 March. Fielding wrote with heavy sarcasm in *The Champion,*

> On Wednesday they began again to work on the new Bridge and notwithstanding the late Alterations which have been made in the Plan, we are confidently assur'd that by the help of an annual Lottery, it will be finished within the Century. At which time it will appear however miraculous and surprising, that certain Persons have converted *Stone* into *Bread.*[19]

The only other casualty of the frost was Samuel Horlock's barge which fully laden with chalk, 'had had the misfortune to run upon the third pier of the Bridge, then unfinished and under water and not guarded by any piles, they having been drawn and drove away by the Frost'. The commissioners ordered Graham to make good the damage and Horlock's barge was repaired for £20 by a shipwright, Thomas Patten.[20]

During the halt between Christmas and the middle of March, the commissioners had arrived at the decision which the general public had imagined to be obvious years earlier, namely that the bridge should be built of stone and not of wood. The piers were of course being built of stone but only with an eye to the possibility of a stone superstructure, and the original decision still held to build the bridge according to James

King's 'curious and extremely well-contriv'd wooden Superstructure' (Plate 12). At a meeting of the board immediately after the New Year, Labelye was called in and asked what he thought would be the difference in time and expense between building the superstructure of timber and of stone. He answered that a stone superstructure would not exceed a timber one by more than £35,000 and could be built in as short a time. Labelye had already put before the commissioners a design for a stone bridge two years earlier (see page 81), and at the end of January he produced the model of one arch of a stone bridge complete with wooden centring, qualifying his estimate of £35,000 which would be exclusive of parapet, cornice, rail and towers, amounting to about another £9,000. He explained his scheme, article by article, at a meeting on 20 February which only thirty-one commissioners attended, resolving because of the magnitute of the decision to postpone it till the next meeting. This was held three days later, forty commissioners being present. Labelye again explained his design, Jelfe and Tufnell agreed that they could build the bridge in two years provided they were not held up by the foundations of the piers, and a motion was then made 'that it is most Eligible that the Superstructure of the intended Bridge should be built of Stone'. The question was put to the vote, twenty going out for the Ayes, twelve staying in for the Noes, and the board then agreed to compensate the carpenters, King and Barnard, for vacating their contract, £2,000 to be paid immediately on account. Labelye was ordered to go ahead with the bridge as high as the top of the balustrade, and in April his appointment was confirmed as

> the Engineer in laying the foundations and conducting the Building of the piers, abutments and superstructure of the said Bridge and all the Works relating thereto.[21]

Labelye's new design, presented to the board on 12 March 1740, was practically the same as the plan he had offered in 1738, differing only in the turrets, cornices, footways and recesses and the decoration of the piers. Many of the improvements had been suggested by Lord Pembroke, 'to whose noble and publick Spirit, constant Care and singular Disinterestedness', says Labelye, 'the Publick chiefly owes, not only the having a Bridge at *Westminster,* but also the having it (perhaps) the best built, and certainly one of the most magnificent Bridges in the whole World'.[22] The new stone bridge was to consist of thirteen semi-circular arches, twelve piers and two abutments, the middle arch to be seventy-six feet wide and the remainder decreasing by four feet each. The design was altered later to include two small arches, one in each abutment, so that the bridge finally consisted of fifteen arches and fourteen piers.[23] The stone to be used was Portland, Purbeck, Cornish moorstone and Kentish rag. A sub-committee consisting of Pembroke, Wade, Arundell and Fox met several times to consider the proper prices to be paid and reported their findings on 26 March.[24]

In the following April and May the commissioners contracted with Jelfe and Tufnell for the three middle arches, four more piers and the two abutments, £4,000 being advanced to them on the security of John Lawson and Edmund Holmes, both soapmakers of St Margaret's, who bound themselves jointly to Ayloffe and Blackerby in the penal sum of £8,000.[25] In June, King's designs for building the centres were recommended by Labelye himself in preference to his own, and King and Barnard then produced an estimate of £2,677 10s od for centres made of five ribs of Riga fir timber. On 11 June the commissioners ordered Labelye to proceed 'with all convenient speed to the driving of the Piles for supporting the Centres of the three middle Arches'. The contract with 'James King, carpenter of St Martin in the Fields, for building and finishing the three middle Centres for turning the Arches', was executed on 30 June, Lord Pembroke offering himself as security for King. A further contract was signed in August for King and Barnard to provide 'a sufficient quantity of good sound oak timber, wrought iron, scaffolding, instruments, &c. for the superstructure'.[26]

The change in plan had of course affected King and Barnard very closely. When the board decided to build the bridge of stone instead of timber the problem immediately arose of how to compensate the carpenters for the release of their contract for the original timber superstructure. At first the commissioners offered 5 per cent of the £28,000 agreed on in 1738 (see page 82). King and Barnard replied that they expected at least 8 per cent. The timber already provided was too large for their own private use, it would be difficult to sell, they were at a loss to know how to estimate their expectation in connection with the timber work needed for a stone bridge, and it would be 'very indecent and even impossible' for them to expect the board to act in any other way than by keeping to the previous contract. The board agreed and £7,472 in compensation was paid to King and Barnard on 9 July 1740: £5,674 for the timber and £1,798 for the 8 per cent on the unexecuted part of the contract. The surplus timber was sold by Graham to Messrs Ransom & Smith, timber merchants at the Stangate.[27]

During the summer the masons worked feverishly at the piers, finishing the fourth pier in twenty days, and by November Labelye was able to report to the board that the sixth pier was nearly finished and the foundations of the seventh about to be dug. The work was made easier by Labelye having asked for a small committee to whom they could refer when commissioners were out of London for the summer.

> As the Hon. Board begins to be but thinly frequented and probably there may very soon be a total vacation of Boards as last year, for the summer season, and as the business of my office will necessarily require frequent instructions and advice, or this part of the undertaking wholly stagnates for that time, I humbly submit whether it may not be proper to appoint a Committee of a certain number of such Comm$^{rs}$ as may happen to be in Town to whom I may have recourse for such instruction and orders.

The board immediately ordered that any three members of the Works Committee should form a quorum.[28]

At the end of the year some slight trouble arose over the ironfounder, Dominic Donelly, charging too much for casting the circular wedges to be used for striking the centres. Donelly asked twenty-eight shillings for 100 wedges. The surveyor said he could find others to cast them at sixteen shillings. Donelly attempted to justify himself before the board by claiming that he had been paid that price for the rollers of the Royal Mint and also for pans used by druggists in melting brimstone. The surveyor then produced another ironfounder, William Bowen, who undertook to cast the wedges for sixteen shillings, 'in as good a manner as those cast by Mr Donelly' who then retired from the interview discomfited.[29]

During the year 1740 the commissioners decided to move their office to larger premises in Old Palace Yard. Their first meetings had been held in the Jerusalem Chamber, a comfortable room with a blazing fire, formerly used by the bishops of the Upper House of Convocation. In July 1737 Ayloffe had been instructed to negotiate for the lease of a house in Duke Street belonging to Samuel Edwards, at the rate of £70 annually for three years. He was also told to buy three dozen chairs and a 'Chair proper for the Chairman of the same sort as those used by the Directors of the Bank of England'. The last meeting in the Jerusalem Chamber took place on 24 August 1737 and then for the next three years the commissioners met in the new Bridge Office in Duke Street.

In March 1740 Ayloffe was told to find a new house, as near both Houses of Parliament as possible, and after examining several in and around Old Palace Yard, he reported to the board that Nathaniel Blackerby's two houses at Parliament Stairs, if laid together, would make a convenient and proper office. A sub-committee then met at the Ordnance Coffee House in Old Palace Yard and decided that the two houses would suit admirably, recommending a rent of £65 a year. The Rate Books show £50 1s 10d paid in Poor Relief.[30] The board resolved to take Blackerby's houses from Michaelmas 1740 provided he made the necessary alterations at his own expense, and notice was given to Samuel Edwards, the landlord of the Duke Street office.[31]

The commissioners' comforts were looked after by a housekeeper, Mary Cundill, who was appointed at a salary of £50 a year out of which she had to provide the necessities to keep the house clean. Mrs Cundill was paid £12 10s 0d quarterly, plus occasional disbursements, from the beginning of 1737 to October 1750, so one assumes she was a treasure giving every satisfaction.[32] The commissioners also employed a sort of handyman or messenger, John Grandpré, who had been with them since June 1736. His salary began at £10 half yearly, then 'having given constant attendance' it was raised in September 1738 to £7 10s 0d quarterly, with occasional £10 gratuities. Payments continued until March 1746 when Grandpré died. He was succeeded by John Lewis, one of the bridge clerks, at a salary of £30 a year, until he too died and the job of messenger was taken over by his brother Thomas.[33]

During the bitter winter of 1739–40 the commissioners presented a petition to Parliament for a Bill to grant them further powers and more money 'that so publick and necessary a Work may not be stopped for want thereof'. The accounts were presented by the treasurer, Blackerby, and showed a deficit amounting to nearly £10,000. Lord Sundon reported from the committee that the commissioners had been confronted with difficulties in cases of litigated titles to various houses in Westminster, especially under *Feme Covert* possessed of estates in their own right and of wives having right of dower. In many cases, he said, hardship might easily arise. Lord Sundon presented the Bill on 18 February and it passed the Commons on 14 March, embodying a clause in answer to a watermen's petition that the compensation money for Sunday ferries should not be paid to the Watermen's Company but to the churchwardens of St Margaret's and St John's. The Bill was agreed to by the Lords on 1 April and received the Royal Assent on 29 April 1740.[34]

The 1740 Act (13 George II, c 16) tried to remove all the disabilities facing the commissioners in treating for land and if the value of land could not be reasonably ascertained the purchase money was to be paid into the Bank of England or the public funds, conveyances to be enrolled at

Westminster. In fact it aimed at granting the commissioners legal power virtually to commandeer property near the bridge approaches, the only redress open to objectors being the right to claim within five years in a book kept by the Middlesex Register. But the difficulties Thomas Lediard and his son encountered during the following seven years show that the new Act was no more successful in this respect than its predecessors.

It authorised a fourth bridge lottery which was to raise £325,000 in sums of £5 or multiples thereof. There were to be 16,510 prizes from £10 to £10,000 with £500 and £1,000 prizes for the first and last tickets drawn, as in the third lottery; 15 per cent was to be deducted for the bridge. The last day for venturing was 15 November and the draw began in Stationers' Hall on 8 December and closed on 19 January, a party of Foot Guards as usual conducting the wheels back to the Lottery Office in Whitehall. The first ticket to be drawn was Number 41,492; a prize of £2,000 was won on the first day by a Mr White at the Plough in Windmill Court, West Smithfield, who shared it with three other venturers; a £5,000 prize was won by Miss Limego, niece to a wealthy West India Jewish merchant of Bury Street, St Mary Axe; one £10,000 prize went to Mr Caldclough, merchant broker, and his partner in the ticket, Mr Thomas, deputy Auditor of the Imprest; a second £10,000 fell to Mr Cappadocia, a Jewish merchant of New Bond Street, who on asking earlier at the Lottery Offices had been told he had drawn a blank.[35]

Fielding's conclusions, expressed in *The Champion,* were 'first that the People are extremely Silly, and secondly that they are extremely Poor'. He attacked the inefficacy of the 1739 Act against deceitful gaming, deplored the activities of the Lottery Office keepers, and suggested that there should be a clause enacted to confine the price of tickets in order to protect thoughtless and simple people from 'a set of Harpies and Vultures still free to plunder Thousands in this public and outrageous manner'.[36]

## NOTES TO CHAPTER 7

1. Bridge Minutes 15 and 22 February 1738/9 and CJ 28 February 1738/9, vol 23 pp 260-1.
2. Bridge Minutes 1 March 1738/9. The almshouses, St Stephen's Hospital, had been founded in 1548 for eight wounded soldiers who each had a room and £5 a year. They were rebuilt in St Anne's Lane in 1741 and renamed the Woolstaple Pensioners (*Notes and Queries* II (1850) p 211).
3. CJ 13 March and 10 April 1739, vol 23 pp 281-2, 325.
4. Act 12 George II, c 33 (1739).
5. *London Daily Post* 7, 8, 15, 18, 19, and 30 January 1739/40, and *Daily Post* 18 January 1739/40. The scheme of the 1739 lottery is set out in *Gentleman's Magazine* IX p 329.

6. Act 12 George II, c 28 (1739).

7. *The Champion* 15 November 1739.

8. Works Minutes 1 August 1739.

9. Bridge Minutes 4 June 1739.

10. Bridge Contracts (15–16) 23 May and 19 June 1739.

11. Bridge Minutes 16 June 1738, 22 January and 12 December 1739.

12. *London and Country Journal* January-February 1739/40, *Political State* 59 January-February 1739/40, *Gentleman's Magazine* X p 35; William Andrews *Famous Frosts and Frost Fairs* (1887) pp 44–51. For the Empress Anna's ice palace see Mina Curtiss in *History Today* February 1973.

13. *The Champion* 12 and 15 January 1740, *Daily Post* 14–15 January 1740.

14. *Political State* 59 (1740) pp 81–2, 94–5 and 100, and most daily newspapers.

15. *Daily Post* 1 and 14 January and 14 February 1740, *The Champion* 3 January, *London Daily Post* 6 February 1740.

16. *Daily Post* 22 January and 18 and 29 February 1740.

17. *The Champion* 8 January 1740. Labelye *Description* p 12 says a third pier was finished before the winter but this is not borne out by the Bridge Minutes.

18. Bridge Minutes 23 and 27 February 1739/40.

19. *The Champion* 15 March 1740.

20. Bridge Minutes 21 May 1740; Bridge Accounts p 43.

21. Bridge Minutes 2 and 31 January, 20 and 27 February, and 23 April 1740.

22. Labelye *Description* p 14.

23. For details of measurements see Appendix page 283.

24. Prices were 2s 3¼d per cubic foot of Purbeck, 2s 11d for Portland, and 17s 6d per rod of rubble (Works Minutes 26 March 1740).

25. Bridge Contracts with Jelfe & Tufnell, 22 June 1738 for the two middle piers, 23 May 1739 for abutments and half piers, 19 June 1739 for four piers, 28 January 1739/40 for six piers, and 2 April 1740 for the three middle arches; Bridge Minutes 2 April 1740.

26. Bridge Minutes 11 June 1740; Bridge Contracts (28–30) 30 June and 13 August 1740.

27. Bridge Minutes 31 January 1739/40; Bridge Accounts August 1740 p 21; *The Champion* 8 March 1740.

28. Bridge Minutes 18 May 1740.

29. Ibid 17 December 1740 and 14 January 1741.

30. One of the houses had been let at £30 a year to Lord Perceval, afterwards Earl of Egmont (Rate Books, St John's ledgers from 1741); *Hist. MSS Comm. Egmont Diary I* 22 June 1732.

31. Bridge Minutes 21 July 1737 and 10 April 1740.

32. Ibid 15 March 1737/8; Bridge Accounts pp 18 and 49.

33. Bridge Minutes 22 June 1736, 24 August 1737, 8 and 15 April 1746, and 28 June 1748; Bridge Accounts p 47.

34. CJ 14 and 25 January, 18 February, 13 and 24 March, 1 and 29 April 1740, vol 23 pp 411–2 and 429–31.

35. *London Daily Post* 9 and 11 December 1740, *Daily Post* 15 December 1740, *London and Country Journal* 20 January 1740/1.

36. *The Champion* 29 December 1739.

# 8
# The Last Bridge Lottery, 1741

In January 1741 the commissioners presented their customary petition to
the House of Commons, beginning with an assurance that they were making progress with Westminster Bridge and with the purchase of property
for the approaches, but 'notwithstanding they have proceeded with the
utmost Care and Frugality . . . find themselves under an indispensable
Necessity of applying to this House for such further Sums as the House
shall judge proper.' Lord Sundon, recounting Lediard's progress report,
told the House that five piers had been finished, a sixth almost finished, the
Westminster abutment finished, the Surrey abutment in great forwardness,
and the middle arch turned fifteen feet high. £59,188 8s 1d had been spent,
all the land necessary for making the first street (from the Bridge to King
Street) had been bought, and the whole, according to Lediard, adding up
to £78,918 3s 7d. The new Bill was presented by Sundon on 12 February.
It was passed on 6 April and received the Royal Assent on 25 April 1741.[1]

The sixth Bridge Act (14 George II, c 40) simply aimed at raising more
money by means of a fifth lottery which was drawn as usual in November
and December at Stationers' Hall. But before the lottery opened a number
of ominous rumblings could be detected in the fortunes of the bridge. In
the first place, all was not well with the firm of carpenters, King and
Barnard, who had contracted for the timberwork the year before. Barnard
was the cause of the trouble, evidently taking offence because his design
for a centre had been rejected in June 1740. He had then failed to deliver
the timber ordered from Guildford and Chertsey and had complained
about his treatment at the hands of the surveyor, Richard Graham. His
case was considered at a meeting of the board in April, the decision being
that his complaints were groundless and 'contained several scandalous
and malicious reflections on the said Richard Graham and is highly
reflecting on the Honour of the Commission'. It was resolved that Barnard
'be never again in any way employed in the service of this Commission'.
Ayloffe was instructed to sue the firm for breach of covenant in the sale
of the superstructure timber.[2] The affair was slightly complicated by a

certain Daniel Chandler, a carpenter employed for about two years as foreman in framing the bottom and sides for the foundations of the piers, claiming that he could do the work considerably cheaper than King and Barnard. However it turned out that Chandler had been dismissed earlier for dishonesty and his claim was judged 'trifling and frivolous and unworthy of any further notice by the Board'. [3]

After Barnard's dismissal King continued, helped by his two sons, James and George. Labelye's certificate, dated 11 June, said that the centre for turning the middle arch had been completed satisfactorily by James King and was ready for the masons. A second centre was finished in September and a third in October. [4] The contract for making and setting up the fourth had been agreed in June and the family continued producing centres, scaffolding and other timberwork until the father died in 1743 and the work was taken over by his foreman, William Etheridge.

Then, early in 1741, a certain amount of criticism was levelled at Labelye alleging that his design was faulty and the workmanship of the masonry inadequate, and that several cracks and settlements in both abutments could clearly be seen. The commissioners immediately ordered two independent surveyors to examine the bridge and James Horne of St James's and Samuel Bond of St Giles in the Fields reported that the charge was exaggerated. They had found one small crack in the wall of the Surrey abutment but judged that it was of little consequence and could easily be repaired, and that both abutments were 'built in a substantial and workmanlike manner'. [5]

During the winter of 1740–1 the weather was severe again, with strong easterly winds and hard frost. [6] It was not as exceptional as the winter before, but bad enough to delay work for several days and, more seriously, to hold up the delivery of stone from the West Country. In fact in February the board was forced to issue a sharp reproof to Jelfe and Tufnell—'the Board doth expect that the Masons employed by this Commission shall always have ready at the Works a sufficient quantity of stone for the carrying on the same without any delay'. Jelfe immediately wrote to his agents at Portland, Thomas Roper and the Tucker brothers of Weymouth, demanding regular cargoes. 'We have received a severe reprimand from the Board', he wrote to Richard Tucker, 'in so much as to threaten dismission if we dont keep up a constant supply of stone'. [7] A few days later Jelfe was able to soothe the board with two letters from Roper, who had already loaded two ships, one with 140 tons of stone, the other with eighty tons. However no stone had arrived by 4 March, the surveyor complained again, and the board 'expressed their great dislike of the want of a sufficient quantity of stone being brought into the River'. Jelfe assured them that he had five vessels loaded with stone—some at Portland, others at sea—and that 'the Works should never stand still for want of stone, he and Mr Tufnell intending to have always a sufficient stock ready in London and Westminster'. Soon afterwards the surveyor reported the arrival of a vessel with eighty tons of Portland stone for the seventh pier, another of 100 tons of Purbeck stone, and two hoys of Kentish rag for the Surrey abutment. And by 15 April 600 tons of Portland stone for the seventh pier and arches had been landed at Westminster and three more ships with 120 tons each for the middle arch had sailed from Portland.[8]

However the supply was still not fast enough to keep pace and Jelfe was compelled to write again, this time to the government surveyors and keepers of the quarries at Portland, Thomas Bryer and Edward Tizard. He reinforced his complaint with a letter to the Prime Minister declaring that unless a high authority intervened, the dispute might not be decided for four months. It would mean a total stop to Westminster Bridge at the best time of year, 'thereby disappointing the impatience and great expectations of the whole town and cause a great clamour among the Commissioners'. Walpole replied immediately with an order signed by his secretary, Henry Legge,

> By Sir Robert Walpole's order, I am to acquaint the Gentlemen who have the stone quarries granted to them that he does absolutely and peremptorily insist upon their complying punctually with the promises they made to him at the time of granting them.
> Westminster 28 April 1741[9]

John Smeaton in his account of the Eddystone lighthouse has left a graphic description of the Portland quarries.[10] He visited various Dorset and Devon quarries in 1756 and was shown round the Portland quarries by Thomas Roper himself, 'a very plain sensible intelligent person'. Roper was at that time quarry manager for John Tucker of Weymouth who had taken him on after Westminster Bridge had been finished. He had been sent down to Portland in the first place by Jelfe and Tufnell to act as foreman of a gang of masons engaged from London. They were to cut the stone according to drawings in order to save freightage on waste. Roper had remained there ever since and when Smeaton arrived to choose stone for the lighthouse, he had had nearly twenty years' experience of the Portland quarries and was able to act as an extremely well-informed guide. He showed Smeaton the local skill of splitting the stone and shaping it with the kevel, a tool with a hollowed hammer at one end and a pick at the other; and the strange carriages used for carting the stone from quarry to pier, 'a kind of cart, consisting of nothing more than a pair of very strong solid low wheels about a yard in diameter, and a very thick axle-tree upon which is fixed a stout planking or platform that terminates in a draught-tree for steerage and yoking the cattle to'. As most of the quarries were on the landward side of the island which sloped gently down to the sea, it was an easy matter to drag the carts to the edge by oxen or horses, then run them downhill to the pier, a block of stone on a sledge acting as a brake.

Smeaton was deeply impressed with Roper's description of the Portland islanders' marriage customs. 'The mode of courtship here', said Roper, 'is that a young woman never admits of the serious addresses of a young man but on supposition of a thorough probation. When she becomes with child she tells her mother; the mother tells her father; her father tells his father, and he tells his son that it is the proper time to be married. But suppose, Mr Roper, she does *not* prove to be with child what happens then? Do they live together without marriage? or if they separate is not this such an imputation upon her as to prevent her getting another suitor? The case is thus managed, answered my friend: if the woman does not prove with child, after a competent time of courtship, they conclude they are not destined by Providence for each other; they therefore separate and as it is an established maxim, which the Portland women observe with great strictness, never to admit a plurality of lovers at one time, their honour is in no way tarnished; she just as soon (after the affair is declared to be broken off) gets another suitor as if she had been left a widow, or that nothing had ever happened, but that she had remained an immaculate virgin. —But pray, Sir, did nothing particular happen upon your men coming down from London? Yes, says he, our men were much struck and mightily pleased with the facility of the Portland ladies, and it was not long before several of the women proved with child: but the men being called upon to marry them, this part of the lesson they were uninstructed in: and on their refusal the Portland women arose to stone them out of the island; insomuch that those few who did not chuse to take their sweethearts for *better or for worse*, after so fair a trial, were in reality obliged to decamp; and on this occasion some few bastards were born; but since then matters have gone on according to the ancient custom.'[11]

In addition to all these obstacles to the advancement of Westminster Bridge—the Barnard affair, the cracked abutment, the harsh winters, the distractions at Portland—the delay in the delivery of stone was increased by the mortal fear of impressment among the crews of hoys and coasters.

War had been declared against Spain in October 1739 but by March 1740 out of thirty ships reserved to protect the Channel only twenty were manned and the Admiralty laid an embargo on all shipping except coasters. The pressmen were not too particular, however, and if they came on a boat manned with three or four healthy seamen, hugging the coasts from Dorset to the Thames estuary, they were unlikely to hesitate. As Labelye reported to the commissioners, 'they dare not sail when there is an extraordinary Press of Seamen for the necessary Service of the Government'. Sometimes a convoy system was used, but this, though insuring a certain amount of safety, also served to increase the delay. There was only one harbour at Portland and ships could be bottled up for days, occasionally weeks, waiting either for a convoy or for the interminable east winds to subside.

Somehow or other they managed to keep the supply of stone going through the summer of 1741. But by October there was trouble again and Jelfe had to overcome, as tactfully as he could, a local quarrel between the government surveyors, Bryer and Tizard, and the quarry proprietor, Richard Tucker. Eventually Tizard came to London to explain matters in person.

> Yesterday (says Jelfe) Mr Tizard called upon me and inquiring the reason of our not having a better supply of stone for the Bridge, find that it is wholly owing to the Malice and ill-nature of Mr Tucker who will not ship the stone when it is drawn down to the Pier but lets it lye there purely to obstruct the business—therefore I charge you peremptorily in case of any neglect whatsoever on Mr Tucker's side, that you immediately order Messrs Tizard and Bryer's people to ship it off directly.

Roper acted fairly effectively, for a few days later Jelfe congratulated him: 'I am glad to find that upon Mr Tucker's hoymen's neglect you made use of Mr Bryer's hoys which proceedings may make Mr Tucker's people more diligent.'[12] A few weeks later Tucker died and Captain Tarzell and Mr Wherry were appointed 'Joint Surveyors of His Majesty's Stone Quarries in the Isle of Portland...a Place of considerable Profit.'[13]

The new management worked well enough and a constant supply of stone flowed into the wharves at Westminster. Pehr Kalm was much struck by the use of Portland stone everywhere, not only for great buildings such as St Paul's, Westminster Abbey and the larger houses, and of course those 'wonderful bridges, London Bridge and Westminster Bridge', but also kerb stones, milestones, gravestones, window frames, coping stones for garden walls, and posts along the sides of the streets to prevent coachmen and carters from driving on the pavements. He described the curious fishy smell it has when it is cut and the exact method of sawing. There were masons' yards all over London and he delighted to watch the men at work cutting the stone with toothless saws.

> The operation is thus: they take the sand which is found near London and sift it tolerably fine. Then it is blended with wet clay laid on a board above the stone they intend to saw, the board sloping towards the sawcut. At the upper end of the board is a bucket full of water. This has a little pipe at the bottom through which the water runs softly across the board carrying a little sand into the sawcut as the saw requires . . . Meanwhile the workmen draw the saw backwards and forwards, the sand continually falling under the blade and thus performing the same service as teeth. They said the reason why they do not use toothed saws is that the teeth would bite so hard and fast that no one would be able to keep up the sawing.[14]

By the end of 1741 it was clear that the bridge lotteries were not capable of financing Westminster Bridge without substantial help from the government. The balance at the Bank of England in favour of the commission was £53,484 14s od. But Labelye's estimate had been £90,000 and in fact when the bridge was finished in 1750 it was calculated to have cost £218,800. The lottery authorised by the 1741 Bridge Act aimed at £275,000 with the usual prizes and penalties and the usual 15 per cent for the bridge. It had originated with a scheme put before the commissioners by Richard Shergold, a lottery broker of Exchange Alley, who said that he had spoken to several large subscribers and found that they had all disliked the £6 tickets because of the size of the loss on blanks. He proposed a lottery of 55,000 £5 tickets (41,875 blanks), with two prizes of £10,000, three of £5,000, ten of £2,000, twenty of £1,000, forty of £500, 200 of £100, 500 of £50, 1,000 of £20, 11,350 of £10, and the first ticket to be drawn to win £500, the last £1,000.[15] Shergold's proposals were accepted, the Bank of England agreed on a payment of £2,000 for their trouble, and the draw began on 23 November 1741 at Stationers' Hall, closing on 29 December.

The first ticket to be drawn won the £500 prize for a waterman of Richmond who had bought it at Hazard's Lottery Office. Another waterman, of Ratcliffe, won another £500, also with a ticket registered at Hazard's. A £2,000 prize was won by Mr Sidsurf, a gentleman of Antigua lately settled in Craven Street. One £5,000 was won by Alderman Benn and another £5,000 by Sir Edward Deering, a Knight of the Shire for Kent. A £10,000 prize fell to Theodore Cock, a draper in partnership with Alderman Arnold of Cheapside. Another £10,000 ticket, sold and registered at Wilson's Oldest Lottery Office, belonged to the Reverend Mr Exton, chaplain to Lord Lymington.[16]

16. Henry, 9th Earl of Pembroke. (*Marble bust by Roubiliac from the collection of the Earl of Pembroke at Wilton House. Photograph by the Greater London Council*)

17. Thomas Lediard (Senior), Surveyor and Agent to the Commissioners. (*Frontispiece to his* Naval History. *Photograph by courtesy of the National Portrait Gallery*)

That was the last of the bridge lotteries. From that time the work was financed by parliamentary grants, purely as a matter of practical economics. The state lotteries went on for another sixty-seven years financing the Sinking Fund and on one occasion, in 1753, helping to buy the collection of Sir Hans Sloane and the Harleian collection of manuscripts. It also built a repository to house them—the British Museum. But the general attitude to the state lottery was changing. Early in the century both Steele and Addison had ridiculed in the *Tatler* and *Spectator* the folly and futility of gambling, though as it happened Addison himself won a £1,000 prize in a lottery of 1711, 'The Adventure of One and a Half Million'. [17] Two of Hogarth's earliest prints satirised gambling—'The South-Sea Scheme' and 'The Lottery', both of 1721—though Hogarth was not above disposing of his own works by means of auctions closely resembling the illegal private lotteries. In fact, the evils of lotteries were patent to most people but few could resist the temptation when such alluring baits were dangled in front of their noses. The shock caused by the South Sea Bubble forced the government to legislate temporarily against excessive gambling but at the same time authorising the company to dispose of its effects by lottery in order to pay its debts. And in 1739 an Act 'for the more effectual preventing of excessive and deceitful gaming' tried to control abuses in betting on the games of Ace of Hearts, Pharoah, Basset and Hazard. It also tried to check the rash of bubble lotteries or raffles selling everything under the sun: houses, land, ships, silver, and even the presentation to Church livings. They had become generally accepted as part of everyday life. [18] The Bill was brought in by Thomas Carew, General Wade and others, and received the Royal Assent on 14 June 1739, but was virtually ineffective, lotteries, raffles and the like continuing to flourish like weeds. A writer in the *Gentleman's Magazine,* using a rather heavy-handed form of satire, advocated a lottery to dispose of 'the distress'd Virgins of Great Britain between the ages of fifteen and forty'. Another writer reported the case of a gentleman's servant who had shot himself after losing all his money and most of his master's. 'Is there a country under heaven but this', he asked, 'where such a House as that in Covent Garden, which has produced more Robberies, Self-Murders, notorious Frauds, fallacious Bankruptcies, &c. than all the Bawdy Houses in the Kingdom, should stand safe after a special Act of Parliament which the World in general thought pointed for its Destruction? No, certainly there is not.' [19]

The national passion for gambling reached its apogee in the 1770s when it became nothing less than a mania, everyone either taking a chance or attempting to cash in on his neighbour's craziness. Tailors, dressmakers, hatters, glovers, snuff and pigtail merchants, shoe-blacks, all entered the game and offered chances. Lottery barbers promised a shave for three-pence with a chance of a £10 win. Chop-houses offered a plate of beef for sixpence and the chance of sixty guineas. Oyster stalls sold a dozen oysters for threepence and a gamble of five guineas. There was even a *Little Lottery Book for Children* (1768) 'containing a new method of playing them into a knowledge of the Letters and Figures'; and Bewick's *New Lottery Book of Birds and Beasts for Children to Learn their Letters by as soon as they can Speak* was published in 1771 with each page in the shape of a lottery ticket and adorned with rather crude woodcuts of Aviary, Ass, Bull, Bullfinch and so on to Buzzard and Zebra, engraved by the young Thomas Bewick, then aged barely eighteen.

The opposition was by no means feeble. A number of satirical engravings appeared in the print shops following Hogarth's 'Lottery' of 1721. 'In Place', a print of 1738 directed against Walpole, showed Captain Jenkins holding out his severed ear to the Prime Minister who sits at a desk covered with papers, one marked 'Scheme for a New Bridge'; another of 1739, 'The State Lottery', shows the Blue Coat boys drawing from the lottery wheels in the Stationers' Hall (Plate 8); another by Charles Mosley shows Harlequin in disguise holding a £10,000 prize beside a lottery wheel with Westminster Bridge floating in the clouds near the more likely end of the adventurers, the King's Bench prison. 'The Humours of the Lottery', an engraving by Parr from a drawing by Gravelot published in December 1740, shows various scenes of good and bad luck in Exchange Alley and the curious behaviour of the winners and losers.[20]

Then in 1737, the year of the first and second bridge lotteries, an article called 'Loss and Disadvantage of the Bridge Lottery and Chance of a Ticket' appeared in *Common-Sense,* written by a man declaring himself a well-wisher to the bridge but an enemy of the ticket touts who 'imposed upon the Credulity of the Unwary by false Representations'. This was quickly followed by 'a very extraordinary Paper written in a very sharp Stile' published in the *Political State* for 1737. It was headed 'Naked Truth: being an exact and impartial Examination of the present *Bridge-Lottery,* address'd to all Persons of common Sense and common Honesty'. Both articles demonstrated how the odds were heavily loaded against the venturers and compared the bridge lottery to 'the Iniquity of the South Sea Scheme'. After analysing the scheme the writer of 'Naked Truth' asked ironically, 'is it to be supposed a single Person among the number at the other end of the Town, for whose Advantage or Pleasure only the intended Bridge is to be built, will stand the drawing of a Ticket of what they have subscribed for, if they can get them off at any tolerable rate? Such a Lottery could never have been filled but with a View to gull and draw in the Country and City of London, if they will be such Fools, in effect, to build a Bridge for them'.[21]

These were statistical criticisms, powerful but impersonal. A more vivid pen appeared against the state lotteries, that of Henry Fielding, who was to succeed De Veil as the magistrate for Westminster. Fielding's testing grounds for the great social satires *Tom Jones* and *Jonathan Wild* were his newspapers, especially *The Champion,* from which he inveighed ceaselessly during 1739 and 1740 against the injustices of the state lotteries and the misery they brought to countless families. He had already written a farce, *The Lottery,* which appeared frequently at Drury Lane, the last and most popular scene actually bringing a lottery wheel on to the stage and showing the dishonesty of the draw. But *The Champion* enabled him, through the mouths of Captain Hercules Vinegar and his omniscient family, to reach a wider public. It was published three times a week and though the articles were unsigned they bear the unmistakable stamp of the master. As they are not readily accessible, a specimen in the form of an ill-spelt letter may be of interest here. It is addressed to Captain Hercules Vinegar of *The Champion* and purports to be from a footman in London to his mistress in the country.[22]

For Mrs Ealce Peartree living with Squire Booser at Hogs Norton in Somersetshire.

Dear Ealce,
    Hopping that you are wel as I ham at present ritin, this cum for to let you no that Mr *Fifa* Atturney wass mistakun about the Lutturi when he zad that twas dree to one but that we lost our Muny because that there were dree Blanks to a Prize. Now, I have vound out a Man that zells all Prizes and nu Blanks and I ave boght twenty vortieth Pearts of twenty Tickets and one may get by one £250 so that by the Whol one may get £5000 vor I ave cast it up but mayhap zum o' um may not cum up zu great Prizes zu that it may not happen to bee above half zu much. Nu Buddy can tell yet, howsumdever, I would ave you enquire of Mr *Fifa* whether that little varm is zuold yet or nu vor I must ave verri bade Lock if I dunt get enuff to bi that. Nu Buddy can tell yet. *Tom Wilson* has got a vortieth Peart of ten Pound and he swears he is out of Pucket but you no, Dear Ealce, there be zum volk that wull never be contented. *Meary Bearns* and *John Haycock* had a whul Ticket betwixt um and this is a cum up a Blank, but they did unt bi un of the seam Man as I dud. I wish you a mery *Christmas* and a happy new Yere, and a grete manny. I wuld zend you zumthing to remeber me but I haf lade out all that I ham worth in the Lutturi and wass vorced to zell mi Zilver Wash into the Bar-

gain. Dunt vere, my Dere Ealce, that Muny shall ever make me valse harted vor if I get the two Ten Thousand Pounds and the dree vive thousands, and the two dree thousands and the dree two thousands, and that I have bin tuld is nut impossible. I dunt mean the whul Tickets but the vortieth Peart o' um. I will give it ale to thee vor if I was to be meade the gratest Squire in ale the Wuld, I shuld never be hapy without my Der *Ealce*. Zu with Zarvice to ale Frinds and Love to Brother *Joo* and Zister *Betty* and Veather I rest.

*My Dear* Ealce's *true Lover till Death*
JHON BULLUCK

*Postscrip*. As zun as I gut but one o the ten thousand Pounds I intend to give Measter warning.

To Fielding and his readers in *The Champion* and many other newspapers of the time, it is clear that the evil lay not in the state lottery itself, which inherently was quite harmless. It afforded an outlet to the healthy gambling instincts of the nation and at the same time provided a useful source of income to the Exchequer. The system of administration was at fault. The evil was concealed in the method of drawing, the overcrowded conditions and in particular the network of insurance, jobbing, division of tickets into fractions, and even blatant forgery, run by the lottery offices to ensnare a gullible public and to line their own pockets. A shilling pamphlet, *The Lottery display'd or the Adventurer's Guide,* tried to protect the public by explaining exactly how the lottery worked. It described the whole process from the appointment of the managers, the printing of tickets, the scheme of prizes, the drawing by the Blue Coat boys, the registering of names, down to the distribution of prizes and the recovery of prizes which had never been received because of mistakes or lost tickets. Mistakes frequently occurred, usually because the clerks in the hurry and confusion had mis-heard the number proclaimed.

> For first with Respect to the Drawing (says the pamphlet), let any person who understands Figures reflect with what degree of accuracy he could enter 14 or 15 hundred numbers (for so many are drawn in a Day) named by a Speaker at the Distance of 40 or 50 feet, and pronounc'd perhaps in a strange Dialect, whilst probably his own ears are dinn'd with the Noise and Clamour of a turbulent Mob behind him.

After the numbers had been drawn, a registered letter was sent off to the adventurer: 'Sir, the Ticket No.—— was this Day drawn a Blank', and the recipient cursed the lottery and threw his ticket in the fire. 'What cruelty!' says the author of the pamphlet, 'what Barbarity!'[23] It is not surprising that quantities of unclaimed prize money remained in the Bank of England.

Tickets usually cost £5, £6, sometimes as much as £10 each, increasing in price as the draw began until on the morning of the last day a single ticket was worth fifty guineas. These sums were of course far beyond the pockets of most people, and the market was operated for them by lottery offices in the neighbourhood of the Guildhall or Stationers' Hall where the draws took place. Brokers bought up blocks of tickets and resold them in fractions, usually in sixteenths. They insured venturers against blanks, blank and prize, and prize only; or they could operate on open receipts even before the tickets were printed—a form known as Light Horse. They would sell them again as Heavy Horse after a rise, sometimes making thirteen or fourteen times their original money. The best-known brokers advertised extensively in the papers. Frank Wilson at his Oldest Lottery Office at Charing Cross directly behind 'the King on Horseback where such Persons who don't chuse to risque much Money have now an opportunity (which may never offer again) to gain the following large Prizes for small sums . . .' Messrs Lowe & Company of Stationers' Alley kept their numerical book in the correctest manner at great expense. Charles Corbett kept his Correct Lottery Office opposite St Dunstan's in Fleet Street where 'Gentlemen and Ladies may be assur'd of having for sixpence each ticket a certain account of their Success immediately'. Thomas Cox, bookseller at the Lamb under Royal Exchange, claimed to prevent frauds by endorsing tickets with his own hand. Richard Shergold, printer to the Bridge Commissioners, operated from the Union Coffee House, Cornhill, moving in 1741 to Pope's Head Alley where he issued a book listing the prizes at the end of each draw. Hazard's Authentic Lottery Office opposite Stationers' Hall reassured the public from a bookshop kept by the Hazard family for forty years. The shop sign over the doorway showed Fortune hovering in the clouds and scattering prizes to the hungry mouths beneath. The brokers are shown in prints of the time—stout well-fed figures leering craftily and holding money bags. Fielding looked on them all as harpies and vultures and the lottery itself as 'the Mongril Progeny of Craft and Fortune, midwived into the World by Jews, Brokers and Foreigners, who have already bespoke the job on the old Presumption that there are Fools enough to pay them well for their Pains'. [24]

During the ninety years between the first bridge lottery in 1736 and the last authorised state lottery in 1826, no less than 151 state lotteries took place, after 1800 sometimes four or five being drawn every year. Eventually a committee of the House of Commons was set up to examine state lotteries in general. Their report, published in 1808, is full of poignant stories of how a gullible public, buoyed up by the illusory hope of something for nothing, could bring itself to destitution and misery. They interviewed lottery inspectors, magistrates, clergymen, pawnbrokers, a treasurer of parish poor rates, two police officers and about a dozen lottery office keepers including a representative of the firm of Hazard & Co still going strong after nearly a hundred years in business.

The committee condemned the various practices inseparable from lotteries and decided that the whole foundation of the state lottery was radically vicious. 'Under no system of regulations which can be devised,' they said, 'will it be possible for Parliament to adopt it as an efficient source of Revenue, and at the same time divest it of all the Evils and Calamities of which it has hitherto proved a baneful source.' They concluded by pointing out that not only did the money accruing to the state fail to compensate for the vice and misery which seemed to be the inseparable companions of lotteries, but when the increase in Poor Rate and the diminished consumption of excisable goods were taken into consideration, the economic advantages of state lotteries were greater in appearance than in reality. It ends: 'No mode of raising money appears to your Committee so burthensome, so pernicious and so unproductive; no species of adventure is known where the infatuation is more powerful, lasting and destructive.'[25] These words were written during the Regency, when gambling at White's, Crockford's and other clubs was at its peak, and it is not surprising that legislation to curb the lotteries was not passed until 1826.

Meanwhile in 1741 the series of bridge lotteries came to an end after five years during which 'the Bridge of Fools' or 'the Lottery Bridge' had contributed its share towards one of the most vicious evils of eighteenth-century life.

A further but luckily not very severe blow to the fortunes of Westminster Bridge occurred towards the end of 1741 when the two Members of Parliament for Westminster, Admiral Wager and Lord Sundon, lost their seats. The occasion was one of the political scandals of the century. Parliament had been dissolved in April and a new election was held in May, the fate of Walpole's unpopular adminstration hanging in the balance. The hustings for the Westminster election were in front of the portico of Covent Garden church, Admiral Vernon and a rich landowner, Charles Edwin, opposing the sitting Members. While the poll was still in progress but slightly in favour of Wager and Sundon, a party of Vernon-Edwin supporters was seen approaching to record their votes. The High Bailiff, John Lever, a Walpole man determined that Wager and Sundon should be elected, immediately produced 'a great number of loose and dissolute Persons furnish'd with Clubs and Bludgeons'. Charles Edwin protested and Sundon demanded that the High Bailiff close the poll books at once and summon the military. To avoid a riot Lever did close the books under cover of a body of fifty Foot Guards with drums beating. They were authorised by three of the Justices of the Peace, Nathaniel Blackerby, George Howard and Thomas Lediard. Wager and Sundon were then declared elected. Luckily for himself Wager was away at the time escorting the King to Holland, but the crowd was so angry that Sundon had to take refuge in the church for four hours and then drive home at full gallop pursued by a mob 'hooping and hallowing, cursing and flinging stones, by which the windows were broke, plenty of dirt thrown into him, one of his footmen's skull cracked by a brickbat, and his lordship wounded in the hand'. [26] The Grand Jury of Middlesex immediately presented a petition to the Court of King's Bench at Westminster, protesting at the conduct of the election and the gross infringement of their liberties. Amid rejoicing throughout the City and Liberty the House of Commons declared the election null and void and ordered the High Bailiff into the custody of the Serjeant at Arms. The three magistrates, Blackerby, Howard and Lediard, were reprimanded on their knees by Speaker Onslow who repeated the Commons resolution that

> the Presence of a regular Body of armed Soldiers at an Election of Members to serve in Parliament, is an Infringement of the Liberties of the Subject. . . . Did you make use of those Powers the Law has invested you with as Civil Magistrates for the Preservation of the Public Peace? No—you deserted all that; and wantonly, I hope inadvertently, resorted to that Force . . .

After the reprimand they were forgiven and dismissed with that 'Sense of Gratitude you owe to the House for the gentle Treatment you have met with on this occasion'. The unhappy John Lever petitioned that he was sorry and wished to be forgiven because his trade as a common brewer was suffering to the ruin of his wife and six children. He too was reprimanded at the Bar of the House and dismissed. [27]

The outcome of the Westminster election was that Viscount Perceval and Charles Edwin were nominated without opposition in December, Admiral Vernon—then at the height of his popularity after Cartagena—having already taken his seat for Ipswich. [28] Wager, aged seventy-five and (according to Horace Walpole) looking like Lazarus risen from the grave, died the following year. He had been a faithful champion of the new bridge from its earliest days, in constant attendance at commissioners' meetings and frequently serving as chairman. Sundon was re-elected as Member for Plympton and continued to serve on the Bridge Commission for many years. He had been the Member for Westminster since 1727 and had done much to find money for paving the streets and repairing the Abbey in addition to his work for Westminster Bridge. But he was one of Walpole's placemen and the manner of his election was quite enough to inflame a sizable mob and to explain the electorate's ingratitude for his fourteen years of labour on their behalf. Besides he was covetous and was said to have owed his peerage to his wife Charlotte Clayton's being one of the queen's favourite Women of the Bedchamber. [29]

## NOTES TO CHAPTER 8

1. CJ 9 and 20 January 1740/1, vol 23 pp 574–5 and 600–1.
2. Bridge Minutes 2 April 1741.
3. Ibid 18 March 1740/1.
4. Works Minutes 17 June, 7 September and 28 October 1741.
5. Bridge Minutes 25 March 1740/1.
6. The newspapers were full of the usual hardship stories; for example, 'a fine black eagle, near three ells between the extremity of each wing was taken in the frost near Northampton' *London and Country Journal* 3 February 1741.
7. Bridge Minutes 5 February 1740/1; Jelfe's Letter Book 7 February 1741.
8. Bridge Minutes 4 and 15 March, 15 and 29 April 1741.
9. Jelfe's Letter Book f 30. *Calendar of Treasury Books and Papers* (1739–41) p 457.
10. John Smeaton *Eddystone Lighthouse* (1793) pp 61–7; see also Rev John Hutchins *History of Dorset* (1774) I pp 586–7, A. M. Wallis 'Portland Stone Quarries' in *Proceedings of Dorset Natural History and Antiquities Field Club* (1891) pp 187–94, and VCH *Dorset* (1908) pp 331–44.
11. Smeaton op cit p 65.
12. Letters to Roper in Jelfe's Letter Book 22 and 27 October 1741.
13. *London and Country Journal* 8 December 1741.
14. Pehr Kalm *Visit to England* 25 April 1748.
15. Bridge Minutes 5 February, 3 June and 11 November 1741.
16. *Daily Post* 27 November, 10 and 30 December 1741, and *London Daily Post* 17 and 25 December 1741.
17. *Tatler* 24 January 1711 and *Spectator* 9 October 1711.
18. Act 12 George II, c 28 (1739); CJ 26 March and 30 May 1739; *Political State* 55 (1738) p 393, and 57 (1739) pp 322 and 345–8.
19. *Gentleman's Magazine* IX (1739) p 149; *Political State* 59 (1740) p 89.
20. British Museum *Catalogue of Political and Personal Satires*, numbers 1730, 2350, 2435, 2446 and 2461.
21. *Common Sense* 25 June 1737 quoted in *Gentleman's Magazine* VII (1737) pp 367–9; 'Naked Truth' in *Political State* 54 (1737) pp 249–52.

22. *The Champion* 3 January 1740.
23. *The Lottery Display'd, or the Adventurer's Guide* (1771) pp 18–20.
24. *The Champion* 1 April 1740. Lottery Office advertisements appear in all the newspapers, especially during or near the periods of the draw, and the brokers were careful to inform the world at whose office the lucky tickets had been drawn (Fielding *Tom Jones* Book 2 Chapter 1).
25. *Report on Lotteries* 1808 (PP 1808 II 182 and 323).
26. *Hist. MSS Comm. Egmont Diary III* pp 219–20 and 233.
27. CJ 9 and 22 December 1741, 23 and 25 January 1741/2, vol 24 pp 54–5. The secret committee inquiring into Walpole's conduct of affairs reported that Lever had received £1,500 at the time of the Westminster election from funds issued under the head of 'Money to reimburse Expences for his Majesty's Service' (CJ 30 June 1742).
28. Admiral Vernon's celebrated order to water down the naval ration of neat rum into 'grog' had been issued the previous year.
29. *Complete Peerage* XIII pp 490–1. For the Westminster election see *Political State* 61 (1741) pp 277–87, British Museum *Catalogue of Political and Personal Satires* 3i numbers 2497–8, 2501, 2505, and George Rudé *Hanoverian London 1714–1808* (1971) p 159.

# 9
# Westminster Bridge, 1742–3

The Westminster election was undoubtedly a shocking specimen of political chicanery resulting, very properly, in the unseating of the two Members. A wider outcome, however, was the resignation of Sir Robert Walpole and the continuation of the war by a government under Carteret and the Pelham brothers. Walpole had led the country for twenty-one years which, except for the last two, had been a period of peaceful consolidation of the Hanoverian succession, free from imbroglios abroad and devoted to establishing the country on a sound economic basis. He was essentially a man of peace. When war was declared in 1739 he said to the Duke of Newcastle, 'it is your war and I wish you joy of it'. Enterprises such as building Houghton Hall and the two bridges at Fulham and Westminster, although of relatively parochial significance compared to the great issues at stake during the term of his administration, interested him profoundly and appealed to his countryman's eye to a developing system of communications. Walpole had been an advocate of Fulham Bridge from the early 1720s, the middle lock actually being named after him, and he had always been sympathetic to the idea of a bridge at Westminster. He had supported the 'Society of Gentlemen' in their struggle to bring a Bill before Parliament, and as a Bridge Commissioner himself occasionally lent his weight to the board in the practical details of the works. He had attended at least four of the early meetings including the first in the Jerusalem Chamber in 1736. Probably it was only age and exhaustion that prevented him from being present after his creation as Earl of Orford in 1742.

Walpole's resignation was not allowed to hinder the progress of Westminster Bridge and a few days later the commissioners petitioned the House of Commons for a new Act authorising the purchase of land in St Margaret's Lane between Old and New Palace Yards—and also, of course, for more money. Lord Perceval, replacing Sundon as spokesman for the commissioners, reported to the Commons from the committee which had examined the bridge officers' accounts of the year's achievements. Thomas

Lediard had said the street from the Westminster abutment to King Street had been laid out and would soon be finished. Ayloffe had reported that the commissioners had spent the sum of £39,868 11s 10d, bringing the total up to £119,257 11s 10d. Labelye had said that ten piers had now been finished, the foundation of the eleventh almost finished, the middle arch entirely turned with Portland stone, the two adjoining arches turned to thirty feet high on both sides, two more turned to fourteen feet, and the abutments on both the Westminster and Surrey shores filled up and finished except for the pavements. The accounts showed a deficit of £12,957 10s 3½d.[1]

Perceval presented the Bill on 26 May. It was passed on 17 June and received the Royal Assent on 15 July 1742. The new Act (15 George II, c 26) authorised the commissioners to buy up land in order to widen St Margaret's Lane and to dispose of any property already bought that should not be wanted. The lottery was replaced with a grant of £20,000, the commissioners to account to Parliament annually. Ayloffe applied to the Treasury for the money to be paid to the commissioners' treasurer, Samuel Seddon, who was forced to report in August that he had been told at the Treasury that no orders had been given for issuing any money. Accordingly a memorial was presented 'begging leave to acquaint your Lordships that the moneys now due to Workmen and others amount to very near the sum remaining in the hands of the Commissioners and that several other large sums are now growing due for works performing by contracts and otherwise, so that unless the sum of £10,000 be now issued by your Lordships order to Mr Seddon, your Memorialists' Treasurer, the Works of the Bridge will very soon be at a stop'.[2]

Samuel Seddon had succeeded Blackerby, who had died on 21 April 1742. He was appointed in June, his securities being John Ludby of German Street and John Crook, perfumer of Pall Mall, at £2,000 each. His salary of £120 a year was paid quarterly until 22 December 1750. Blackerby's death was one of a number of misfortunes besetting the bridge at about this time. He had been the commissioners' treasurer since 1736 (see page 68) and died owing them over £2,000 which caused a good deal of concern to his son Benjamin and to his fellow magistrate and executor, Richard Farwell, who had been a member of the Society of Gentlemen in 1733 and one of Blackerby's guarantors in 1736. Their lawyer tried to pacify the board with a sale of the Blackerby property in Westminster but this took too long and Ayloffe was instructed to summon another of the securities, Sir George Caswall, to attend the Bridge Office for payment. Sir George meanwhile had also died but his son George came and was given until 18 January 1744 to pay the debt, later extended to 1 March with the chance to confess a judgement to the commissioners with a Release of Errors on the Bond. On the advice of his solicitor Caswall refused and Ayloffe was ordered to bring an action. A curious situation then arose. Ayloffe applied to the Duke of Somerset, an absurdly arrogant old man of eighty-two but the senior Bridge Commissioner, asking him to return Blackerby's bond, made payable to the duke and other commissioners, and to provide a letter of attorney so that he could bring an action. The duke declined at first, alleging that he had been nominated to the Bridge Commission without his consent, had never taken part in its proceedings and would have to take the oaths before he could act as a commissioner. Ayloffe was able to disillusion him at least on the last point. He waited on him in person, braving the peremptory treatment menials expected from 'the proud Duke', and eventually persuaded him to deliver up both bond and letter of attorney. The case was ordered to be proved in Chancery as John Schellinger et al. versus Benjamin and Anne Blackerby et al., the amount claimed being reduced to £1,352 15s 2d mainly because of three years' rent due to the estate from the Bridge Office lease. It came up again in 1749 when Schellinger, acting for the commissioners, sued Anne Blackerby who had inherited from her father the office of housekeeper at the Palace of Westminster. The question at issue was whether a fee of six shillings and eightpence a day ('an allowance made by the Lord Chamberlain's warrant to Blackerby for the purpose of making clean and sweeping the Royal House and Palace and the chimneys, removing and placing the chairs, forms, &c. in the House of Lords') should be held as part of the profits of the office. The Lord Chancellor ruled that it would be very extraordinary if this should not be subject to the creditors' satisfaction and the Blackerbys lost their case. Anne however, with her sister, continued to act as housekeeper till her death in 1780 in her apartments adjoining the House of Lords.[3]

Because of frosts and east winds the masons were unable to fulfil their contract to complete the middle arch by the New Year, but on Wednesday 3 February, a few days before Walpole's resignation, the middle arch of Westminster Bridge was turned, decorated with flags and streamers, and visited by several of the commissioners and other distinguished persons.[4] The timber centre beneath the arch was not actually struck for another two years but the arch itself was there, its brand new Portland and Purbeck stone shining proudly in the middle of the river for all to see. Perhaps there was some truth in Fielding's unkind comment that the masons had been ordered to 'turn the grand Arch of the *Lottery-Bridge* as fast as possible— that the People may have something to stare at for their Money'.[5]

During the course of the year six piers on the Surrey side were finished and work continued on the two arches on either side of the middle arch. Contracts were signed for the remaining four arches next the Westminster abutment and for the torus, cornice, towers and footway over the middle arch.[6] The bridge at this stage was painted by Samuel Scott, probably in the spring or early summer of 1742 (Plates 20-2).

The works went ahead at a good speed and on the whole, in spite of the formidable tidal rise and fall, accidents were relatively few. When they did happen, the commissioners dealt with them in a humane and expeditious way. For instance in 1739, when the piers were being built, Robert Pealing, a carpenter employed on the scaffolding, had his thigh broken by a falling baulk of timber. He petitioned the board for relief and was immediately awarded £10 'in consideration of the great damage and sufferings which he received in the service of the Commissioners', and was appointed doorkeeper and watchmen on the Westminster abutment at a wage of nine shillings a week. Unfortunately as a result of this generous allowance to Pealing, the doorkeeper and watchman on the Surrey abutment, who had been enjoying two shillings a day (Sundays excepted) was reduced to nine shillings a week too. Pealing had to be dismissed after four years because he 'had been of late very remiss in his Duty and behav'd so very ill as to be of little use to the Comm$^{rs}$.'[7]

In 1742 Sarah Clarke petitioned the board for relief because her husband Thomas Clarke, a labourer, had died from an accident at the bridge. He had been 'assisting in craning up a large piece of timber and the tackle giving way unfortunately fell upon him by which means he received several Bruises of which he languished 5 days and then died, leaving the Petitioner poor, aged and infirm'. Sarah said that her husband's death left her wholly destitute and the commissioners resolved to give her £10 'in consideration of her calamitous circumstances'.[8]

Accidents of another sort were also liable to occur and to plague the commissioners with litigation which dragged on sometimes for years. Barges coming down stream occasionally lost control, either by mistake or deliberately, and ran foul of the bridge works. This was especially likely to happen at high tide when submerged piles could become a danger to navigation of which some watermen were swift to take advantage. The barges were enormous, carrying cargoes of over 100 tons, usually of malt and meal, and could certainly be difficult to manage on a strong ebb tide.

The Vessels which bring this Malt and Meal to Queen Hithe (says Defoe) are worth the Observation of any Stranger that understands such things. They are remarkable for the length of the Vessel and the Burthen they carry and yet the little water they draw; in a word some of those Barges carry above a Thousand Quarter of Malt at a time and yet do not draw Two Foot of Water . . . Some of these Barges come from Abingdon which is above one hundred and fifty miles from London if we measure by River.[9]

Horace Walpole said they looked as solemn as barons of the Exchequer as they passed under his window at Twickenham.

One such accident happened to a barge owned by a timber merchant of Reading, John Abery, who claimed that on 25 November 1741 one of his barges carrying timber to Deptford had been sunk while passing between two piers of the bridge. He blamed it on the lack of defence piles put up by the bridge officers and not on any want of skill or care on the part of the men sailing the barge. The board went into the matter and decided that 'the accident was entirely owing to the fault and misconduct of the Bargemen (Joseph Vicars, Edward Rivers and John Perry), and that the Barge came down in a very hard gale of wind at which time it was very improper for any Barge to be broke loose. Notwithstanding which the Bargemen might have steered right thro between the two piers which were then above water if they had taken common care'. They decided that Abery was not entitled to relief and though he appealed again a few weeks later, the board stood firm.[10]

On the other hand a claim was presented by George Field who said that as he was coming down the river on 3 December 1741 with a barge loaded with tiles, hay and faggots, he struck one of the piers, then under water, and the barge sank. Graham and Labelye were called in by the board and both agreed that it was true that the ninth pier had been under water then and that they had not been able to replace the fence piles which had been removed the day before to let out the caisson. Field's claim was allowed though it was reduced to £12 from £19, which the board judged to be exorbitant.[11]

These claims were settled fairly quickly, one way or another. A more protracted case was that of Thomas Greenaway, a barge owner of Reading, who laid a petition before the Works Committee saying one of his barges had been sunk on 31 August 1743 by striking an underwater pile and that the cargo had been seriously damaged. The board examined Greenaway and his solicitor, Mr Rochester, who confirmed the accuracy of the schedules and lists attached to the petition. Another inquiry took place a week later, Greenaway giving the names of the men aboard the barge at the time.[12] The board then resolved that it required an affidavit sworn before a Master in Chancery from every individual aboard describing exactly what happened to cause the accident. Greenaway petitioned again in April 1744, the board instructing Ayloffe to lay the case before their solicitor, Taylor White, who advised that the commissioners were not liable to make good damages sustained by boats and barges running foul of the necessary bridge works. A further petition was laid before the Works Committee by Samuel Hawtyn, James Allen and James King, agents for the several sufferers who had owned cargo in Greenaway's barge. Eventually Greenaway petitioned Parliament and a clause in the Bridge Act of 1745 authorised the commissioners to compensate him for the damage sustained to his barge and its contents of meal, malt and other goods. The Bridge Accounts for March 1746 record a payment to Greenaway of £302 17s 6d.[13]

Another accident befell barge number 480, belonging to Samuel Anderson of Brentford. According to Graham, 480 was 'wilfully and with design by the Bargeman on board, run against a Pile of the Bridge whereby the said Pile was beat down, and that the Bargeman swore he would beat down another'. Anderson was summoned before the board and claimed that the barge did not belong to him but to Mrs Fawkener, also of Brentford, and had been sailed by William Payne, innkeeper of the Waterman's Arms, Brentford. Payne attended the board in person, admitted his great misdemeanour, and humbly asked for pardon offering to repair the damage at his own expense. The Board accepted and the record was signed WP, 'the mark of Wm Pain'. Payne paid up five months later but only after a special messenger had been sent to Brentford threatening prosecution.[14]

A certain number of the accidents of this sort were undoubtedly caused by vindictive watermen, disgruntled at the whole idea of a bridge jeopardising their livelihood and—as they never tired of repeating—bringing ruin and destitution to their families. They would do anything they could to hinder the works without actually infringing the Act, which specifically declared that anyone maliciously damaging it could be treated as a felon with the penalty of death without benefit of clergy. The innkeeper of the Waterman's Arms narrowly escaped this fate but for the most part the accidents were genuine enough and fairly dealt with by the commissioners. For instance, the bargemaster to the bridge, Samuel Price, petitioned the board because a number of piles had been smashed and the stumps below water remained as a hazard to navigation. One of his lighters, loaded with stone, had been holed and sunk in this way. It had to be pumped out and raised over four tides and between twenty and thirty workmen employed to transfer the stone to other lighters. Price successfully claimed £4 3s 0d for this, and another £6 8s 7d for a shipwright's repairs to his barge.[15]

The case of Searle's wharf was another example of a reasonable settlement between a petitioner and the commissioners. Thomas Searle, the owner of a large wharf at Stangate near the Surrey abutment, petitioned for compensation for loss and damage caused by the caissons, pile-driving engine, barges and other craft moored alongside, especially 'at all times when Frost and very high winds the said Engine and Barges have been haul'd in close to the Wharf causing mud to settle . . . & at all emergencies yr memorialist has been very assisting and serviceable to the Works as is well known to your Hon. Officers'. Searle's petition was referred to the Accounts Committee, Graham agreed that on many occasions he had been very helpful, and they settled amicably on twenty guineas' compensation.[16]

The innkeeper of the Waterman's Arms at Brentford had another reason for disliking Westminster Bridge. Some years earlier a local committee engaged on rebuilding Brentford Bridge had employed Labelye, 'a skilful engineer', to survey the old bridge and prepare a design for a new one of brick and stone. Labelye's design and estimate of £1,200 was approved by the committee in May 1740, but the inhabitants of Brentford vigorously resented the design, which they claimed perpetuated the defects of the old crooked bridge. This old bridge, they said, being 'built in a very crooked manner was exceedingly dangerous and inconvenient to all passengers, and by its narrow turning, frequently occasioned the overthrowing of coaches, chaises, carts, waggons and other carriages'. Labelye's bridge would cause 'imminent danger to the lives of all persons passing over the said Bridge if it be built in the same form in which it is now carrying on'. Their protests were unavailing and the new Brentford Bridge, to Labelye's design, was built by Thomas Dunn and Charles Marquand and finished in March 1742.[17] It was disparaged by Batty Langley who—as usual seizing every opportunity to sling mud at Labelye—said the bridge was of no significance and the river there no more than a ditch.[18]

Among the legal entanglements of this time was the case of Crooke v. Shone. Early in 1742 a petition was presented to the Board by Peter Shone and several other workmen employed by James King, explaining that an action had been brought against them in the Court of Common Pleas by Richard Crooke, sueing them for trespass in his yard at Stangate. They claimed that they had done this on the personal instructions of Labelye but their cause had been so ill-defended at the trial that Crooke had obtained a verdict of £305 damages. The Board indemnified them immediately

18. *Prospect of the Frozen Thames* by Jan Griffier II 1739–40. (*Guildhall Library, City of London. Photograph by J. R. Freeman*)

19. Frost Fair, 1740 – sightseers inspecting the middle piers. (*Engraving published 26 February 1740 by Elizabeth Foster. Photograph by courtesy of the Ashmolean Museum*)

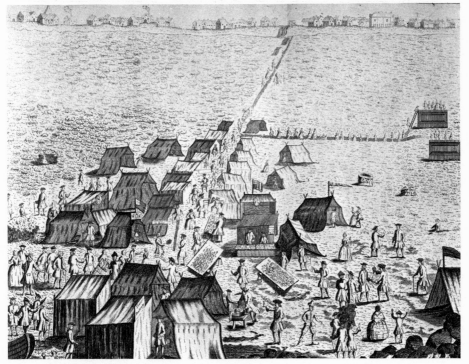

20. *Westminster from the North East by Samuel Scott, 1742. (Bank of England collection. Photograph by courtesy of the Paul Mellon Foundation)*

against all costs and damages, and briefed Taylor White and his junior, George Monkman, to draw up a Bill in Chancery against Crooke on behalf of Shone, Etheridge and others. Several £50 payments were made to Monkman during the course of 1743 and 1744, and the case was heard before the Master of the Rolls on 2 March 1744 and again in December before the Lord Chancellor, but the original verdict in favour of Crooke was upheld on both occasions. Crooke was also awarded £48 costs.[19]

In spite of all the accidents and litigation, at the end of 1742 with the twelve large piers completed and the fourth arch well under way, Labelye felt justified in publishing a sixpenny pamphlet called *The Present State of Westminster Bridge* which was written in the form of a letter to a friend and dated Westminster, 8 December 1742. The friend, possibly Lord Pembroke, had evidently expressed a desire for a description of the bridge and Labelye seized this opportunity to refute his critics who 'speak out of Malice or Ignorance or both'. *The Present State* begins with a short description of the bridge as it was intended by the commissioners, with details of the breadth of the river and measurements of piers and arches. It continues with a rough time schedule from 1736 to the end of 1742, and ends with an account of the works already finished during the three years, eight weeks and three days since Pembroke had laid the first stone in January 1739. The pamphlet was published anonymously but judging from the rather bombastic style already familiar to readers of the *Short Account* of 1739, it was almost certainly written by Labelye himself. Moreover large sections of it were later lifted bodily and incorporated in his final publication, *A Description of Westminster Bridge,* which appeared over his own name in 1751.

Early in January 1743 the commissioners petitioned Parliament for more money. They reported that they had made considerable progress in building the bridge, had purchased several houses and parcels of ground, and had spent all the previous years' grant of £20,000 authorised by the Act of 1742. The contracts already entered into amounted to 'a greater sum than is now in the Hands of the Petitioners . . . who therefore pray the House to take the Premises into Consideration and grant them such further Sums of Money as to the House shall seem meet'. The petition was then referred to a Committee on Supply which recommended that £25,000 be granted to His Majesty to enable the commissioners to finish the bridge. There was no Bridge Act that year.[20]

By the spring of 1743 the masonry had been turned for five arches as can be seen clearly in Scott's paintings. During the year eight timber centres were being used to build the arches from the three middle arches to the

Westminster shore, moorstone steps were laid on the abutments, Robert Smith hoisted and filled ballast up to the level of the keystones on the middle arches, and contracts were signed with Jelfe and Tufnell for building the east 68ft and part of the east 64ft arches, and for the cornices and octagonal towers of the west arches. [21]

At the end of the year the first centre was struck and on 14 December Labelye announced to the commissioners that

> in Obedience to an Order dated 28th September 1743 I do report that the Center under the West 68 ft Arch is now entirely struck down & free from the Arch which remains entire without the least Crack or visible settlement. [22]

This triumphant moment, when the first arch of Westminster Bridge stood alone and open for all to see, was the culmination of months of preparation. Over a year before, a contract had been signed with an ironfounder of Tonbridge, William Bowen, to cast fourteen circular wedges for striking the centres. 'The Centers', ran the contract, 'are to be eased, lowered and struck by means of certain machines or instruments called circular wedges, contrived by Mr James King and to be made of the Toughest & Kindest Cast Iron. . . .' [23]

During the summer the board decided that the five centres under the arches nearest the Westminster shore should be eased and struck and that Labelye should begin immediately with the west 68ft arch [24] Preparations for striking the centre began in September, scaffolding was erected, and levers and circular wedges were assembled. But the method did not prove as efficient as had been expected and on 30 November Labelye reported to to the board that should King's machines prove ineffective he had 'other ways for discharging the Centers, with or without the use of Circular Wedges, which having been already examined by proper persons, have been thought as safe and expeditious as any other'. The board then ordered the surveyor to provide the necessary levers so that work could begin as soon as possible. [25] On 5 October Labelye reported that work had started and that King had made four long booms to defend the arch while the centre was being struck, the long delay being caused by blacksmith's work (twenty pairs of screwed staples and a pattern hook for the levers) not being delivered. This was ready in a fortnight and on 19 October Labelye reported that five circular wedges had been placed between the oak plates of the centre. A week later twelve out of the twenty were in position and the rest were put in as fast as tides and weather would permit.

By 2 November all wedges were in place and several of the blockings from between the oak plates had been cut out. The centre was struck on 14 December 1743 and the arch stood free, the contract with Jelfe and Tufnell having been signed three and a half years before, in April 1740. [26]

The problem of striking the centres involved the board in another quarrel with James King who at the critical moment became difficult and refused to provide timber to make the levers for turning the circular wedges. Labelye wrote several letters telling him to start work striking the first centre but King preferred his own method and forbade his workmen to co-operate in any way, so that Labelye was compelled to suspend him from his post as carpenter of the bridge works. William Etheridge, King's foreman, was then asked whether he would obey Labelye's orders. He assured the board that 'he always had been and was now willing and ready to obey Mr Labelye's orders and to be paid by the Commissioners for his work'. It was resolved that Etheridge be employed as bridge carpenter till further orders. King deeply resented his suspension but it is evident that his health had been affected and he was probably slightly mad. The disappointment of his timber bridge being rejected in favour of Labelye's stone bridge, the difficulty he found in working under Labelye's orders, the failure of his pet invention the circular wedges, and the mental and physical strain of battling to erect the centres in severe and unfamiliar conditions, had proved too much for him. He wrote a letter to Labelye in Billingsgate language. Then he wrote another letter to the board trying to justify himself. [27] He pleaded that the 'distemper brought upon me by my close attendance on the Works of the Bridge' was being spitefully twisted to imply that he was 'touched in the head'.

The board was moved by this letter from a man who had served faithfully from the beginning, and King was reinstated on condition that he and his workmen obeyed Labelye and that Etheridge should continue as foreman as usual. [28] The glad news was brought to his bedside by Richard Graham who said that King and his family had received him with tears of joy. But a few months later Labelye complained again that King refused to obey his orders and the board suspended him once more, reappointing Etheridge as master carpenter. King replied that he did not know what his crimes were and Ayloffe was told to tell him that he had been suspended 'for disobeying orders given in relation to the carrying on of the Bridge Works'. This was the final blow. King died a few weeks afterwards and Etheridge became bridge carpenter in his place.

In the summer of 1743 the bridge suffered another loss in the death of its surveyor and agent, Thomas Lediard, who had worked assiduously on the commissioners' behalf since the early days of the Horn Tavern meetings and the Society of Gentlemen. On 8 June the chairman of the Bridge Committee read out a letter from Lediard

> purporting that his Ill state of health at present is such that he cannot possibly attend his Duty at the Bridge Office or abroad so well as he could wish, and he believes his Son is by his directions capable of every Branch of it, is desirous that he might surrender his place & that his Son might be appointed to succeed him, & in which case he would himself do all the Business that can be done at home & would as soon as his health permitted attend the other part of his Duty as usual.

The Board then resolved to appoint Thomas Lediard the younger to the office of surveyor of the ways, streets and passages and agent for buying and selling such estates as the commissioners shall find necessary . . . at £300 p.a. in the room of his father who had resigned. Thomas Lediard the elder was then appointed assistant surveyor and agent without salary.[29]

Lediard made his will on 11 June 1743, 'being weak in body but sound of mind and understanding'. By virtue of his being a citizen of London he bequeathed his estate to his wife, Helena, and the residue to be divided amongst his children, Thomas, Elizabeth, Dorothy, Sophia and Jane, share and share alike. His son Thomas was also given his gold watch and seals, but another daughter, Helen, and her husband William Gardiner, peruke-maker of Threadneedle Street, were given 'the sum of one shilling each and no more'. He died on 14 June unrepentant of this last piece of irony, and the will was proved on 16 June before the worshipful William Strahan, Doctor of Laws Surrogate, Thomas Lediard and James Horne being the executors.[30]

The part Lediard and his son played in the development of Westminster has been very little recognised by students of London topography and is more fully treated later in this book. From the days when the first Bridge Act was being prepared for Parliament until 1750 when the bridge was completed and the roads on either side were in busy use, the two Lediards had devoted themselves to the problem of clearing and re-planning the approaches to the bridge in both Westminster and Lambeth, and the lay-out of eighteenth-century Westminster, especially from New Palace Yard to Whitehall, owes a great deal to their efforts.

# NOTES TO CHAPTER 9

1. CJ 3 and 25 February 1741/2, vol 24 pp 82 and 94-5.
2. Bridge Minutes 4 August 1742.
3. The Blackerby affair came up before the board at frequent intervals in 1742-5 and was recorded in the Minutes. A final account appeared in the Minutes for 2 April 1745. For the 1749 judgement see Francis Vesey *Reports in Chancery* I p 347 and Belt's *Supplement* pp 172-6.
4. *London Magazine* (1742) p 151.
5. *The Champion* 17 April 1740.
6. Bridge Contracts (50) 2 April 1742, Jelfe & Tufnell 'for the remaining four arches on the Westminster shore' and (48) with James King 'for building and finishing four Centers for turning the Arches'. (53) 30 June 1742 with Jelfe & Tufnell for torus, cornice, towers and footway over the middle arch.
7. Bridge Minutes 30 April 1740; Works Minutes 26 September 1744.
8. Bridge Minutes 3 November 1742.
9. Defoe *Tour* (1725) p 143.
10. Bridge Minutes 24 February 1741/2.
11. Bridge Accounts April 1742 p 45.
12. The bargemen were John Harding, John Perry, Robert Pink, Edward Townshend, Robert Parker, Thomas Alexander, William Gibbs, Jonathan Brooks, John Webb, and one other whose name Greenaway could not remember.
13. Greenaway's case is recorded in the Bridge Minutes and Works Minutes throughout 1743-6, Bridge Accounts 1746 p 78, CJ 6 March 1744/5, and Act 18 George II c 29 (1745).
14. Bridge Minutes 6 June 1744, Works Minutes 12 September and 3 October 1744.
15. Bridge Minutes 8 January 1744/5.
16. Bridge Minutes 23 May and 13 June 1744.
17. Middlesex Session Books, May 1740, May and June 1741 (Orders of Court IVff 171, 211d-212d, 217). Michael Robbins *Middlesex* (1953), p 77.
18. Anonymous (but probably by Batty Langley) *Observations on a Pamphlet . . . by Charles Marquand* (1749) p 6 (see Bibliography page 300).
19. The Crooke-Shone affair is recorded in the Bridge Minutes of 1742-4, especially 2 March and 19 December 1744; the case does not seem to have been reported.
20. CJ 10 and 27 January 1742/3, vol 24 pp 370-1 and 396.
21. Bridge Contracts (61 and 64) with Jelfe & Tufnell, 31 August and 23 November 1743.
22. Bridge Minutes 14 December 1743. Graham's congratulatory letter to Pembroke is in the Wilton Papers.
23. Bridge Contracts (56) 23 August 1742.
24. Works Minutes 31 August 1743.
25. Bridge Minutes 30 November 1743.

26. Works Minutes October-November 1743, pp 197–203.
27. Bridge Minutes 7 and 14 December 1743 and 14, 21 March and 6 June 1744.
28. Ibid 6 June 1744. King died in April 1744.
29. Ibid 8 June 1743.
30. Westminster Wills filed in Act Book 3–12, number 2404W (Westminster City Records).

# 10
# Westminster Bridge, 1744-5

The Bridge Minutes for the year 1744 open with a resolve to set before Parliament a petition explaining the commissioners' doubts and difficulties over the purchase of land. The petition was duly drawn up and printed as usual by the official printers, John and Robert Baskett. In February Lord Perceval, reporting from the committee which had examined Ayloffe on the problem, declared to the Commons that the previous Acts were useless when it came to the point of producing evidence for the titles of estates after valuation by a jury. It was plain that the commissioners were not allowed enough time and were too severely restricted.[1]

The actual valuation by juries presented little difficulty. Frequently Ayloffe, or more usually Lediard, was unable to arrive at any conclusion with the negotiations for conveyances, and the commissioners were driven to empanel a jury. The usual procedure was for Ayloffe to issue a precept to the bailiff of the Dean and Chapter of Westminster Abbey directing him to enrol and return a jury on a certain date, normally about three weeks ahead, to assess the value of the property in question, or the damage and re-compense to be awarded. On the agreed date the bailiff attended at the door of the Bridge Office. He was called in, and read over the names of the jurors he had summoned. The jurors were then sworn in and informed of the occasion of their being convened and of the powers vested in them by the Bridge Act of 1736. This authorised the commissioners to treat their verdict as 'binding to all Intents and Purposes whatsoever'. The bailiffs were paid fifteen shillings for summoning each jury, and the deputy bailiff was paid ten shillings. Between 1740 and 1750, the deputy bailiff, Samuel Baldwin, was paid £87 12s od and on one occasion an honorarium of £28 15s od for 'his extraordinary care and attendance' in summoning twenty-five juries.

The evidence for both sides was examined, and the jury withdrew and either came back with a verdict or went off to look at the premises under consideration. They usually returned a verdict as directed by the com-missioners. Sometimes jurymen failed to answer the bailiff's summons and were penalised with a £5 fine, but this was easy enough to evade, especially if the culprit was a man of standing. For instance on 16 January 1740 sixteen jurymen failed to turn up, among them Sir Philip Meadows of Whitehall, James Joy of Duke Street, Sir Andrew Chadwick of Broad Street, James Kendal and Thomas Patterson of Conduit Street and Sir Thomas Clarges of George Street. They were all ordered to be fined £5 each and a relief jury was sworn. At the following meeting fourteen appeared in person with reasons for non-attendance and their fines were remitted. Sir Thomas Clarges was excused his fine a fortnight later, his reason being that he had been out of London on the day of the summons. Thomas Patterson had to pay the fine, the only unlucky one of the sixteen. [2]

Jurymen were treated to free dinners at the Crown Tavern. At first these cost about £6 but they were not slow to seize a good chance and by 1744 the dinners were costing about £20. On one occasion, in July 1743, no doubt to help them decide a particularly troublesome case, the bill came to £21 3s 8d and the Accounts Committee decided that the inn-keeper of the Crown, Mr Lewis, should be restrained, as 'some indulgence to the palate both in Meat and Drink seems to have been rather more liberal than necessary'. [3]

After the juries had brought in their verdicts, the commissioners' agent and surveyor, Thomas Lediard, or in some cases Ayloffe himself, was then faced with the task of unravelling the titles of estates, many of them mortgaged or leased and sub-leased several times. The aim of the new Bill was to make the earlier Acts more effective in this respect.

Lord Perceval presented the Bill to the Commons on 13 March, together with a petition from the watermen of St Margaret's and St John's praying that the Sunday ferry compensation should not be paid to the Watermen's Company, where it would be swallowed up and never seen again, but into the hands of their own stewards. A clause was inserted about the exchange of tickets unclaimed from the lotteries. The Bill was passed on 17 April and received the Royal Assent on 12 May 1744. [4]

The eighth Bridge Act gave the commissioners powers to interview anyone whom they believed to hold title deeds, books, papers or writings, in order to help them discover the ownership of land and make their task of purchase easier. The verdicts were to be sent to the Clerk of the Peace who was to endorse the date on the back and send them on to the Surrey Quarter Sessions to be kept among the records. The commissioners were then considered to be in possession. [5] The Act defined the ownership of the bridge in case 'any Treasons, Murders, Felonies, Riots, Assaults or other Misdeeds' should be committed upon it. From the centre of the middle arch to Bridge Street was to be part of the parish of St Margaret's, and from the centre to the eastern abutment part of the parish of Lambeth, provided that later on it should not become a county bridge. Any further land purchased by the archbishop was to be recorded in the diocesan registry and approved by the Dean and Chapter. The Company of Watermen was required to pay the Sunday ferry recompense into the hands of stewards chosen by the watermen of the parishes of St Margaret and St John the Evangelist. The sum of £25,000 was granted for the bridge from the supplies vote. And arrangements were legalised for the disposal of unclaimed lottery tickets.

Meanwhile the bridge was beginning to look more substantial than the ghostly illusion appearing in the satirical engravings of a year or two before. In January 1744 Labelye showed the board a drawing for the centres to turn the abutment arches, making use of the frames on which the caissons had been built and thus dispensing with the need to buy more timber. The board approved and the centre on the Surrey side was set up soon afterwards. The two centres for the east 68ft and 64ft arches were also ordered to be set up making use of the centres from the corresponding west arches.[6] The three middle arches had long been turned, the fourteenth and last pier was finished on 21 February and Labelye's certificate for the completion of the four arches nearest the Westminster shore (exclusive of the cornice and balustrade) was dated 20 March. As Labelye proudly announced in his *Description*

> The building of both Abutments and all the Piers of this Bridge were happily compleated in Five Years & twenty-three Days from the laying of the first Stone, notwithstanding all the Stops and Difficulties occasioned by the Tides, bad Weather, Ice and frequent Wants of Stone which was kept from us by long easterly Winds, besides some Embargoes, extraordinary Pressing of Seamen, and staying often for Convoy in Time of War.[7]

During the summer, work began on the centres for the eastern arches and by October the arches themselves were beginning to show and the cornice and footway over the middle arch were finished. A great deal of work was also done towards laying out the position of the approach road over Lambeth Marsh. Contracts were signed with Jelfe and Tufnell for building the next two eastern arches and for the balustrade over the middle arch according to drawings and a model shown by Labelye to the Works Committee in February. Jelfe and Tufnell's securities on this occasion were Barwell Smith, 'a gentleman of large estate in the County of Surrey (Esher) and possessed of a considerable place in the Exchequer', and

Francis Tregagle of New Inn, London, 'reputed by everybody that know him as a Gent of a plentiful fortune'. Jelfe and Tufnell were allowed six months to finish each arch from the time the centres were fit for use.[8]

In August an imposing line engraving by Fourdrinier was published from a drawing by John Maurer but it shows the new bridge complete even to the last lampholder over the balustrade. It was made either from Labelye's model of the bridge or from one of the drawings by his foreman, Thomas Gayfere. The engraving, with modifications to bring the lamp-holders and other embellishments into line with the final design, was used a few years later as an illustration in a new edition of Stow's *Survey of London*.[9] A more accurate and certainly more pleasing idea of the state of the bridge in the summer and autumn of 1744 can be seen in Richard Wilson's painting in the Philadelphia Museum of Art, showing a view from the south and with the bridge half finished.

Once the first centre had been struck it was necessary to remove the piles on which it had been built in order to prevent them becoming a further navigational hazard to be pounced on by watermen and bargees, anxious as ever to discomfort the commissioners. Most of the guard piles protecting the piers had already been drawn; in 1742 Labelye estimated that the number of long piles left was about 280 and short piles 290. Two good second-hand lighters were used for extracting them by means of the tidal rise and fall, this method being cheaper than cutting them off close to the river-bed.[10]

During the severe winter of 1742-3 Graham complained to the board that one of the iron-shod piles had been drawn by the frost and floated upstream to Chelsea 'where it had been taken up by one Norris, a gardener, who had refused to deliver it'. Ayloffe was ordered to write to Norris to say that unless he gave it up the board would prosecute. Norris brought back the pile immediately.[11] But a more scientific method was needed and Labelye reported that in his opinion the cheapest and safest way of disposing of the short piles on which the centres were built would be not to draw them out but to saw them off as low as possible below the natural bed of the river. James King, two other carpenters, and 'an experienced workman' were then asked to tender and the contract was given to Charles Marquand who agreed to cut off all piles at ten shillings each, beginning with the west 68ft centre and completing each arch in a month. Marquand agreed to provide at his own expense all 'saws, Boats, Barges, Scaffolding, Tools, Instruments and Utenzills whatsoever', and in exchange to take for his own use all piles over five feet long at sixpence per cubic foot.[12] The contract was signed on 27 February. Unfortunately Marquand was 'seized with a violent fit of the Gout' and unable to complete the first centre in the time limit. He was given another fortnight but the work was not even begun, the contract was cancelled and Labelye ordered to cut down the piles himself. Marquand's failure did not cause much surprise. He was a Guernsey man who had settled in Westminster and in 1737 had sent in a plan for a timber bridge which 'was considered and laid aside as insufficient'. He had built Brentford Bridge successfully enough but his scheme for pile-cutting was ridiculed by the bridge opponents. Batty Langley declared that the model 'which he said his Servant in his absence had lighted the Fire with—I do prophecy it will be of little use to himself or the Publick.'[13]

After trying out various methods, Labelye magnanimously reported that Etheridge had invented an engine better than three of his own design and several others. It was able to cut eight piles every tide and by June the west 68ft arch had been cleared.[14] Etheridge's pile-cutting engine was a simple contraption consisting of a horizontal saw fitted into a frame running to and fro on brass rollers. It could be lowered deep into the water, held in position by a ratchet lever with a sharp point which dug into the pile, and worked by two men hauling ropes on a floating platform (Plate 11).

Ominous rumours began to circulate during the summer and autumn of 1744. People said that all was not as it should be with Westminster Bridge and in September Labelye was asked by the Works Committee his opinion of 'the common report of the Town that there were cracks in some of the Arches'. Labelye replied that he knew of no cracks in any of the arches or other works of the bridge, but that what were in public talk called cracks were only open joints. These were only in the east spandrels of the middle arch and the west spandrels of the 52ft arch near the Westminster shore. The Works Committee then ordered him to produce in writing his views on why the cracks or open joints has appeared and what might be their possible consequence. He was also required to report whether he thought the piers would settle further when the centres were struck. Labelye told the committee that the centre under the west 72ft arch could be safely struck and that the carpenters could be properly employed on this work immediately.[15]

Both Labelye and Jelfe and Tufnell handed in their reports a week later and both were careful to distinguish between cracks and open joints, which in the committee's order were treated as the same. Cracks or fractures were defects caused by irregular settlements, too long bearings, false joints or bad workmanship, and none of these, they claimed, existed in the new bridge. Open joints, on the other hand, were a natural accompaniment to the building of arches, and especially bridge arches, which were built on piers not founded on solid rock and must of necessity settle slightly. Furthermore where open centring was used, as in Westminster Bridge, the elasticity in the timber and the increasing weight as the arch rises compresses the lower joints and opens the upper joints until the key is inserted.[16] They all stated firmly that when the centring was struck the open joints would close automatically.

Labelye also reported at some length and after 'long calculations founded not only on Mathematical and Mechanical Principles . . . but also upon Observations and Experiments', concluding that when the centres should be struck none of the piers would be in danger of settlement. He had confined his calculations to the middle arch which had long carried loads of material far in excess of the weight it would eventually be called on to bear. And what could be proved for the middle piers naturally followed with the others. [17]

At the end of the year Labelye struck the centre under the middle arch and in his report to the board declared that the arch became entirely free from the timbers before the centre had been lowered half an inch. The few small open joints in the spandrels were closed even before the centre was eased, 'on account of the Blockings having already given some way by their long sodding, and that after striking the Center the said open joints entirely closed conformable to what I delivered as my opinion in my Report of 12th September last'. The west 52ft centre was struck a few weeks later, the arch becoming free before the timber had been lowered a fraction of an inch, and the open joints closed the moment the centre began to be eased.[18]

The ninth Bridge Act, last of the almost continuous annual series between 1736 and 1745, once again was concerned mainly with the approaches and with acquiring the usual grant from the supply vote. In February 1745 the commissioners petitioned Parliament for powers to sell land with the object of preserving the uniformity and beauty of the approaches. Lord Perceval presented the Bill to the House of Commons on 26 February and it was passed a month later, receiving the Royal Assent in May.[19] The new Act empowered the commissioners to enter any house or ground to examine 'any Nuisance, Annoyance, Obstruction or Inconvenience' and if they decided harm could be caused to the bridge or its approaches they were authorised to remove it immediately. Seven days' notice had to be given to the owner. People causing nuisances could be penalised. £25,000 was then set aside from the Aids and Supplies granted to the king for the year 1745, and the Act concluded by authorising the commissioners to compensate Thomas Greenaway for the damage sustained to his barge and its cargo (see page 140).

Although money for the bridge had been authorised by Act of Parliament it was not so easy to lay hands on it. Early in the year the treasurer told the board that he had been several times to the Treasury to see the secretary, John Scrope, for a letter instructing the tellers of the Exchequer to issue the £15,000 due to the bridge from the grant of £25,000 in the last Act. The board then gave him a memorandum 'praying for the speedy issue of the said sum of £15,000 to Mr Seddon' and threatening to discharge several workmen if the money was withheld any longer. Ayloffe himself applied to the Treasury and was assured that the Commons 'would on Wednesday next be moved for making good the Deficiences of the last Year's grants', and a few days later Seddon informed the board that he had received £15,000 and had deposited it in the Bank of England. Then again in August the Works Committee ordered Seddon to apply immediately to the Lords of the Treasury for their £25,000 and to say that unless the money be issued forthwith a stop must be put to the works of the bridge; £15,000 was received in September.[20]

The reluctance of the Treasury to pay out was understandable considering that the country was at war and funds for civil projects were short. Furthermore the news of the Pretender's landing in Scotland in July (percolating to London a month later) and his progress to Prestonpans, Carlisle and Derby during the autumn was an additional source of alarm. There was a distinct falling off in the attendance of commissioners at meetings in the Bridge Office, only Pembroke and Cavendish regularly taking the chair. Perceval and Edwin, the Members for Westminster, turned up occasionally early in the year but not at all after June. There were no meetings between 23 July and 8 October but this was quite normal, the majority of commissioners invariably being away from London every summer. Labelye was in fact given leave of absence in July to advise the Duke of Bedford on works to be undertaken by the Corporation of the Great Level of the Fens. His two reports on the subject were of some help in overcoming opposition from the townsmen of King's Lynn and Cambridge and in rebuilding the Denver Sluice in 1748–50.[21]

In spite of the Treasury the work went ahead at much the usual speed and early in the year the remaining three centres on the Westminster side were struck and rebuilt on the Surrey side. The successful striking of the centres was due to a change in method. In February 1744 Labelye had reported that the circular wedge method had proved a failure, acting very differently from what had been expected. He had then proposed the use of straight wedges, explaining them with the use of a model and so far impressing the board that they agreed to allow him to build the east 68ft centre with this method of striking in view. Graham was told to provide the necessary timber and to keep an exact account and Labelye was ordered to strike the west 64ft centre on 15 February 1744 using straight wedges.[22] The centre in Canaletto's drawing in the Duke of Northumberland's collection is of Labelye's improved design (Plate 35). During the summer of 1744 both the west 64ft and 60ft centres were struck by this method, and in December the centre under the middle arch was struck, followed a few weeks later by the west 72ft and the centre nearest the Westminster abutment. Labelye reported

> in Obedience to an Order from this Hon Board bearing date 22nd January 1744/45 I have caused the Center under the Western 56 ft Arch to be eased & struck quite clear from the stone work on Fryday morning last. The whole operation was not one hour long in the performance and succeeded perfectly well so that the Arch remains fair and entire, wholly supported by its Piers, as are the Middle Arch and the six Western Arches of the Bridge, of which I thought it my Duty to acquaint the Honble Board being, Their most humble and most Obedt Servt. Chs Labelye. 26th February 1744/5.[23]

At the end of the year Etheridge reminded the commissioners of his two inventions, a battering ram and an underwater saw, which he claimed had saved them a great deal of time and money. Graham and Labelye confirmed this claim. Etheridge's method of striking the centres with straight wedges instead of James King's circular wedges had saved two thirds of the time and expense, and his pile-cutting engine was also much cheaper than Marquand's estimate. In all they calculated that Etheridge had saved the commissioners over £1,500. He was awarded £200 'for the advantage the Works of the Bridge have received and are still receiving by his Ingenious Contrivance of lowering the Centres by means of and on the principle of the Battering Ram, & by his Invention of the Engine now used for cutting the Piles underwater and close to the Bed of the River' (Plate 11).[24]

Once the centres were struck the timber was advertised and sold, the ribs, cheek-pieces and braces to a carpenter, John Phillips, and the puncheons or underworks to Chelsea Waterworks. One of the oak plates used for striking the centres was cut up, fitted with iron rings and fixed to each angle of the two abutments for the purpose of mooring boats to the causeways.[25]

This meant that except for the abutment arch the whole of the West-minster side of the bridge was clear of timber by the spring of 1745 and building activity had been transferred to the eastern arches. The three eastern centres were ordered to be set up in January and contracts were signed with the masons, Jelfe and Tufnell, for building the last two arches (east 56ft and 52ft) together with their upperworks—torus, cornice, plinths, balusters, pedestals, rail and footway. [26] The masons also presented their final account for the three middle arches and parts of the two ad-joining arches contracted for in 1740. Graham reported that he had measured them, Labelye pronounced them satisfactory, and Bowack said that Jelfe and Tufnell had been paid £24,000 in six instalments and that £74 9s 6d was owing. The board ordered this to be paid directly. Labelye's certificate, typical of many, was transcribed into the Minute Book: [27]

> This is to certifie all whom it may concern that Messrs Andrews Jelfe & Saml Tufnell have finished the three middle Arches and part of the Arches next to the middle ones and performd the work in a substantial and work-manlike manner and perfectly agreable to their Contract dated 2nd April 1740, witness my hand Westminster 7 January 1744/5 Chs Labelye.

Certificates for the other western arches had already been signed in March 1744. During the summer and autumn of 1745 the masons hurried on with the eastern arches and by the end of the year three more had been very nearly finished. Between October and December, when the east 72ft and 68ft centres were struck, the bridge must have looked much as it appears in Antonio Jolli's painting, except that the west 68ft centre unaccountably remains under its arch (this was the first centre to be struck, in December 1743) and the masons appear to be still working on the balustrade of the middle arch, whereas this was actually finished in September (Plates 25-7). [28]

Another view of the bridge towards the end of 1745 or early in 1746 is a composite painting by Samuel Scott in the Paul Mellon Collection, probably painted years later from sketches and drawings and even from memory (see pages 232-3). The balustrade in this case has been built as far as the west 68ft arch (Plate 23). The completed cornice and balustrade can be seen in detail in another of Scott's paintings, also in the Mellon Collection (Plate 42); a slightly different version is in the Tate Gallery. Labelye's first design of 1736 shows the parapet decorated over each pier with pairs of stone balls and tall conical lamp-posts. The drawing is lost but the design survives in a copy by one of the masons' clerks, Thomas Gayfere, and in an engraving by Fourdrinier (Plate 5). The final design was issued in 1739 and shows the piers surmounted by the well-known alcoves, the four central and the abutment ones being covered (Plate 6). An alteration to the design, possibly suggested by Pembroke, was decided by the commissioners in 1743. They resolved that the parapet walls should 'consist of four courses of Portland Stone, that the first plinth shall be cramp'd and run with lead at every joint; that the Dado Course shall be joggled 2 inches deep at the least into the second plinth & upper Rail; and that the upper Rail shall be doubly plugg'd with Iron and run with Lead at every joint'. At the next meeting Labelye produced drawings for the cornice and parapet and for paving in the recesses. [29]

A few months later he showed the Works Committee drawings and a model for the balustrade over the middle arch. They were approved by the board and the contract signed with Jelfe and Tufnell in March 1744, to be completed within the space of four months. The cornice over the middle arch was finished towards the end of 1744, the balustrade a year later. The commissioners then inspected the work and resolved 'that the plinth, balustrades, pedestal and rails over all the arches of the Bridge be with all convenient speed built and erected and made in the manner and agreeable to the drawing and model shown to the Commissioners on 22 February 1743/4 by Mr Labelye the Engineer'. [30] The work was finished almost exactly two years later when Labelye's completion certificate was produced for the board: [31]

> This is to whom it may concern that Messrs Jelfe and Tufnell have completed the Cornice, Foot-ways, Parapets & Ballustrade over both the Eastern and Western half of Westminster Bridge in a proper and workmanlike manner and every way conformable to their Contract bearing date 23rd November 1743 and 17th December 1745.
> 20th November 1747                    Chs Labelye.

Most of the building of the two abutments was finished by the end of 1745, though the final touches were not completed till 1747. The contract for the abutments and two half piers had been agreed in May 1739 at a cost of £11,358 15s 0d and 'that the half piers of the Bridge shall with all convenient speed be erected of the best Portland Stone . . . the abutments of the Bridge shall likewise be erected and built with all convenient speed . . . and the masons to provide Tarris, iron for cramps, barrs and chains sufficient to strengthen the said Piers'. [32]

Work on the foundations began immediately the contract had been

21. Detail of Plate 20. The Westminster abutment.

22. Detail of Plate 20. The four middle arches.

23. Westminster Bridge to York Watergate, 1743–50 by Samuel Scott and Canaletto. (*Drawing in the Paul Mellon Collection. Photograph by courtesy of the Paul Mellon Foundation*)

24. Westminster Abbey, 1744, Westminster Bridge, 1745, and the Ironmongers' Company Barge, 1749, by Samuel Scott. (*Paul Mellon Collection. Photograph by courtesy of the Paul Mellon Foundation*)

signed; a contractor, Edward Nicholson, provided labourers at one shilling and sixpence per tide to dig, wheel, pump and spread ballast.[33] Nicholson, the chief manager for sinking the foundations of the piers, was paid continuously from 1739 to 1741. He died at his house near Millbank in March 1741 and was succeeded by Robert Smith, who continued dredging and providing ballast until the bridge and the roads were finished in 1750.[34] By February 1743 the abutments were complete except for the arches and for filling and paving. The last half pier was finished in March 1744 and the commissioners examined a model to help them decide how the end of the bridge should be paved.

> The Westminster abutment of the Bridge shall be forthwith filled and levell'd, the carriage way pitch'd with the best Guernsey pebbles, sorted of not less than 9 inches not more than 12 inches pitch. The plying places pitched with Purbeck Squares not less than 6 inches deep (except the margins next the carriage way which are to be lined with Moor Stone one foot broad and backed with Purbeck), and that a Parapet Wall be erected thereon four feet high or thereabouts agreable to a model this day shown to the Board.[35]

A fortnight later Edward Mist tendered for paving the Westminster abutment, agreeing to keep the pavements in constant repair for two years. His tender was accepted against that of another paviour, John Wilkins. In August 1744 Labelye told the Works Committee that it was absolutely necessary to build the abutment arch on the Westminster side, the centre which had been used for the Surrey abutment being removed and in danger of decaying. The committee ordered the work to be put in hand with all expedition.[36]

The following year Labelye produced drawings for four causeways to run from the abutment steps into the river as far as Low Water Mark, about one hundred feet on the Westminster side, and eighty feet at Lambeth. The causeways were to be sixteen feet broad and set not quite parallel to the bridge itself, and to be founded on the old timber from the four long sides of the two caissons. There was also to be a narrow footway under the two 25ft arches connecting the stairs on either side. The abutments and steps were often slurred over in paintings of the bridge but an accurate picture can be seen in Canaletto's drawing of about 1747 in the British Museum.[37] Periodic payments of £1,000 were authorised to Jelfe and Tufnell up to £10,000 in December 1745, and the masons' final account, £997 18s 0d for work on the lower flight of waterstairs north and south of the Westminster abutment, was dated 18 November 1746.[38]

By this time the contracts had been ordered for the remainder of the superstructure and Labelye felt it his duty to tell the Works Committee that they would be likely to run into heavy expense if they continued to pave the whole footway as they had paved the middle arch—that is to say, with moorstone laid in ballast. He suggested that one guinea a foot (£2,000 and time) could be saved by making the footways of three-inch Purbeck stone with moorstone margins laid in good mortar in a foundation of Maidstone rubble. The moorstone pavements already laid on the middle arch could be taken up and used for the margins.

Meanwhile the paviour, Edward Mist, had died and according to the newspapers, 'Mr Dorset, belonging to His Majesty's Board of Works, is appointed Paviour and Scavenger at Whitehall in the room of Mr Mist, deceas'd'.[39] However, the contract was won by Thomas Phillips, who agreed to pave the Westminster abutment with Guernsey pebbles at five shillings a yard. Phillips continued to act as paviour to the commissioners for the next three or four years, levelling the roads and laying pavements on the bridge itself and the newly laid out Parliament Street. Stone for the 'walking paths' came from Lanlivery near Fowey and was coarser grained than moorstone, with long white spars called 'dogs' teeth'.[40]

## NOTES TO CHAPTER 10

1. Bridge Minutes 25 January 1744; CJ 1 and 27 February 1744, vol 24.
2. Bridge Minutes 16, 23, 28 January and 6 February 1739/40.
3. Ibid 13 June 1744.
4. Act 17 George II, c 32 (1744); CJ 4, 17 April and 12 May 1744.
5. Conveyances are to be found in three volumes in the Public Record Office— Estates purchased, Leases granted 1749–98, and Valuations of properties purchased, precepts, juries' verdicts, etc. 1739–46 (PRO Works 6/60–2).
6. Bridge Minutes 16 and 23 May 1744.
7. Labelye *Description* p 75.
8. Bridge Contracts (72) 30 May 1744.
9. Strype *Survey* (1754–5) II p 641.
10. Works Minutes 22 September 1742.
11. Bridge Minutes 22 December 1742 and 5 January 1742/3.
12. Bridge Contracts (66) with Marquand 27 February 1743/4. Bridge Minutes 18 April 1744.
13. Charles Marquand *Remarks on the Different Construction of Bridges* . . . (1749) together with Batty Langley (?) *Observations on a Pamphlet entitled 'Remarks etc' in which the Puerility of that Performance is considered* (1749). Marquand reappeared in 1749 with this pamphlet describing how Labelye ought to

have built Westminster Bridge and including a hare-brained scheme for transporting ready-built piers up and down stream by means of air-filled floats. This idea was utterly squashed by Batty Langley's sledge-hammer and thereafter Marquand disappears into oblivion.

14. Bridge Minutes 20 June 1744.
15. Works Minutes 5 September 1744.
16. 'Open centres' had no support under the middle of the crown.
17. Works Minutes 12 and 26 September 1744.
18. Bridge Minutes 22 January 1744/5.
19. Act 18 George II, c 29 (1745); CJ 1, 14, 26 February, 6, 25 March and 2 May 1745, vol 24 pp 732 ff.
20. Bridge Minutes 15 January 1744/5; Works Minutes 13 August 1745.
21. Works Minutes 1 June 1745. For Labelye's work in the Fens see Bibliography page 299 and H. C. Darby *The Draining of the Fens* (1940) pp 127-9 and 139-41.
22. Bridge Minutes 1 February 1743/4. Labelye's improved wedges were used to strike the centres of Blackfriars Bridge and can be seen in Piranesi's magnificent engraving published in 1764.
23. Bridge Minutes 26 February 1744/5.
24. Ibid 3 and 10 December 1745.
25. Ibid 8 April 1746 and 13 January 1746/7.
26. Bridge Contracts (69) 21 March 1743/4.
27. Bridge Minutes 15 January 1744/5.
28. Jolli's painting was engraved by Fellows and published in J. T. Smith's *Antiquities of Westminster,* there attributed to Canaletto and in the collection of the Hon Percy Wyndham (Constable *Canaletto* 437).
29. Bridge Minutes 9 and 23 November 1743.
30. Ibid 29 October 1745.
31. Ibid 1 December 1747.
32. Bridge Contracts (15) with Jelfe & Tufnell 23 May 1739.
33. The foreman, Joseph Napper, was paid two shillings a tide.
34. Bridge Accounts pp 39-40; *London Evening Post* 3-6 March 1740/1.
35. Bridge Minutes 21 March 1743/4
36. Works Minutes 22 August 1744.
37. Reproduced in Constable *Canaletto* 752.
38. Works Minutes 19 November and 17 December 1745. The masons' accounts were settled in Works Minutes 13 October 1747. *National Journal* 5-7 June 1746 (Bodleian Library).
39. *Westminster Journal* 28 September 1745.
40. J. Smeaton *Eddystone Lighthouse* (1793) p 51.

# 11
# Westminster Bridge, 1746–7

As had become the custom by now, at the beginning of 1746 the commissioners petitioned Parliament for a further grant towards the bridge and the petition was referred to the Committee on Supply, which voted £25,000.[1]

During the course of the year the works went ahead so well that by 25 October the last stone of the bridge, except for balustrade and pavement, was laid (unfortunately without ceremony) by the Earl of Pembroke, who had 'laid the first stone but seven years, eight months and twenty-seven days before'. The east 64ft and 60ft arches were finished in March, the cornice and parapet walls in April, and on 20 July the last arch was keyed in, so that 'from that Day', says the engineer proudly, '(if it had been proper) the Bridge was passable both for foot Passengers and Horses.'[2] The last two arches, the 56ft and 52ft on the Surrey side, were finished in November 'in a proper and workmanlike manner' according to contract, and their centres were struck and removed in November and December.[3]

Plans for the new road in Surrey were well in advance. Most of the land had been bought and contracts signed with Robert Smith for the supply of Thames ballast for the foundations. In fact the bridge was so nearly finished that the *Gentleman's Magazine* published a supplement to the volume for 1746 containing a detailed description of the whole structure with measurements of piers and arches, the weight of stone, the width of the river and tidal data, and claiming the superiority of Westminster Bridge over the long bridges at Ratisbon, Dresden, Lyons, St Esprit and Madrid, none being 'equal to this either for strength or magnitude, regularity or quantity of water which they cover'. The author remained discreetly silent on the subject of the bridges of antiquity, in comparison with which Westminster Bridge pales into insignificance. The article was embellished with an engraving by Thomas Jeffreys showing the bridge complete except for the domed shelters and lamp-posts, and demonstrating how admirably its sturdy Palladian arches linked the Gothic complex of Westminster Hall and the Abbey to the fields and marshes of Lambeth.[4]

The balustrade was not to be finished for another year but even without it the new bridge in 1746 was a splendid spectacle with all its shining Portland and Purbeck arches complete. Until the winter three timber centres remained near the Surrey shore to add interest and contrast. The fame of this feat of combined architecture and engineering travelled abroad and in the summer of 1746, inspired by eulogies from visitors to Venice, Canaletto himself arrived in England, possibly with the encouragement of the commissioners, eager no doubt to lure the master to try his hand at painting the London scene. Certainly several of them were among his most ardent admirers and patrons. The Duke of Richmond and the Duke of Bedford, both commissioners of the bridge since the first Act of Parliament in 1736, had been collecting Canaletto paintings for many years; and Sir Hugh Smithson (later Duke of Northumberland) and the Duke of Beaufort both became patrons of the artist very shortly after his arrival in this country. It would be tempting to pursue this theme but on the other hand, Europe being fully occupied with the War of the Austrian Succession, and Englishmen being unable to travel on the Grand Tour, Canaletto may well have decided that the slackness of business in Venice could make a visit to his English patrons both timely and profitable. This is the generally accepted view but it seems reasonable to guess that the new bridge provided an extra inducement.

A few months after Canaletto had settled in London—probably in the Duke of Richmond's house near the Westminster abutment—there occurred one of the annual river pageants which made the Thames between the City and Westminster an even brighter and more gorgeous scene than it was normally. On 29 October the Lord Mayor of London, William Benn, visited Westminster to be sworn in office before the barons of the Exchequer. The Lord Mayor in his state barge was accompanied by the aldermen and City officials and the state barges of the various City companies, the whole cortège being rowed upstream by crews of uniformed watermen who prided themselves on their seamanship in negotiating the arches of the new bridge and bringing their heavy craft alongside the old King's Bridge at Westminster. This was an event to which the eye of Canaletto, tuned to the regattas, ceremonies and Ducal festivals of Venice, could do full justice. The 'Lord Mayor's Day' in the Paul Mellon Collection almost certainly represents this occasion in 1746, an engraving by Remegius Parr being published early in 1747. The fine new barge of the Goldsmith's Company—the Livery Company of the previous Lord

Mayor, Sir Richard Hoare—closely accompanies the Lord Mayor's barge, which is shown broadside manoeuvring for position before passing beneath the middle arch (Plates 31-2).

Unfortunately from the point of view of the history of the bridge, Canaletto's painting is of little value. In order to enhance the occasion he has shown the bridge complete, with the balustrade finished, the centres removed, and Labelye's design improved to the extent of fitting each recess with a domed roof instead of only the four at each end and the four at the middle. Furthermore the balustrade over the middle arch is decorated with baroque statues of the two river gods, Thames and Isis—a fanciful embellishment considered at one time but never in fact carried out. However, the painting is a marvel of translucent colour, brightness and gaiety, and no doubt it would be totally insensitive to a great artistic achievement to mention that, from all contemporary accounts, London and especially Westminster after daybreak was almost invariably shrouded in a thick pall of smoke.

The abutments and landing stages were finished by the end of 1746 but before their completion some slight difficulties had to be met from the occupants of the new vaults built under and adjoining the abutments. James King, the works carpenter, had leased the Stangate wharf and taken possession of the vaults beneath. After his death in 1744 his sons, James and George, petitioned the Works Committee to be allowed to retain the vaults, believing that their father had assumed he could have the same privilege on the Surrey abutment as Andrews Jelfe had at Westminster. The committee decided that neither the Kings nor Jelfe had any rights to the vaults and ordered Labelye to fill them up and block the doorways opening on the stairs of the abutments.[5] The bricked-up entrance to Jelfe's vault on the Westminster abutment can be seen in one of Canaletto's drawings in the British Museum.[6]

Several of the bridge officials had fairly early on in the works invested in land adjoining the new bridge. In 1740 Lediard had agreed to let Jelfe have a building lease for seventy-two years of an important area of land between Bridge Street and the north-east corner of New Palace Yard. This estate later became the Stationery Office but at first it was used by Jelfe to build thirteen houses. He had some difficulty in letting them. In fact in 1745 he petitioned the board to remit his ground rent till the bridge should become passable because he had 'laid out near £9000 on the said Ground which has lay dead near three years & as yet produces very little, there being only two houses lett at an under Rent and no prospect of letting the remaining eleven houses till such time as the Bridge be passable'. Three years later he acquired the freehold for £905.[7]

One or two minor accidents happened at this time. William Oldfield, a mason employed on the bridge, 'had the misfortune in removing one of the stones of the octagon Towers to fall down and break his thigh by which accident he was and still is rendered incapable of getting his living & therefore prayed for some Relief'. The board awarded him a compensation of £15.[8] Then one Sunday morning in October a 124ft West Country barge belonging to Mr Cobb of Reading was sunk after striking a submerged lighter under one of the arches. Soon afterwards the papers reported that a soldier, 'being disguised in Liquor, fell from the top by which accident he fractured his skull and was so terribly bruis'd that he was carried to the Infirmary in James Street and is since dead'.[9] John Grandpré had died this year too, in March or April, having served as the commissioners' messenger since 1736. His job was taken on by John Lewis, one of the bridge clerks, at £30 a year.[10]

An accident of a different kind happened in the summer, when a youth of seventeen was drowned while bathing near the Westminster abutment. The bridge was a favourite bathing place (see Plate 42). It was easy to walk down the new causeway into the water; but on this occasion the youth 'ventured a great way into the River though he could not swim, fell into a Ballast hole and was never seen more. Some boats went off in search of him but the body was not found.'[11]

This sort of fatality was very sad but from the point of view of the bridge works a more serious spate of damage to the actual superstructure began at this time. The previous year Labelye had complained to the Works Committee that 'a great number of persons continually climb'd upon the Bridge and hindered the Masons in their work'. The surveyor was ordered to 'cause a Fence Spik'd and Tenterhook'd to be set upon the end of the West 56 ft Arch', and another a few months later was set up, 'spik'd and tenter'd' on the Surrey abutment.[12] These fences were only moderately successful in keeping at bay bands of hooligans from both Lambeth and Westminster, and in June 1746 Labelye reported that 'disorderly persons every day of the week but more particularly on Sundays, got upon the Bridge and did damage to the Works carrying on there'. The Board ordered a constant watch to be kept and the door to be locked on Sundays, the key to be handed to Etheridge.[13] However, the bridge was so nearly finished that it was difficult to keep people off and in September *The Westminster Journal* reported that 'Westminster Bridge is now so near finished that a Coach might pass over it, and the workmen are putting up the Balustrade's coping, part of which is affix'd up already on the Westminster side'. After a few weeks the same newspaper reported that 'a Company of Soldiers march'd over Westminster Bridge with Drums beating which is the first instance of any number of people going over together'.[14] This was directly contrary to the regulations. Labelye complained to the board and notices were printed, stuck on the barriers and published in the papers, forbidding 'Soldiers or other Persons of any Degree whatsoever' from crossing the bridge or hindering the workmen.[15] The notice quoted the Act's clause about malicious damage and declared that offenders lawfully convicted could be found guilty of felony and 'suffer death as a Felon without benefit of Clergy'. Furthermore the board applied to the officer commanding the Guards for a posse of soldiers to be posted at the abutments in case of any disturbance.[16]

At the beginning of 1747 the Commons voted £30,000 from Supplies to enable the commissioners to carry on with the bridge.[17] Parliament was dissolved in June and the new Parliament met in November, with Viscount Trentham and Admiral Sir Peter Warren as the new Members for Westminster instead of Perceval and Edwin. Sir Hugh Smithson was returned as Member for Middlesex.

The trial of Lord Lovat in Westminster Hall ended in the sentence of death and his departure to a barbarous execution on Tower Hill. 'God bless you all, and I bid you an everlasting farewell', he said to his judges as he left Westminster Hall, 'we shall never meet again in the same place, I am sure of that'. The sentence, pronounced on 18 March, must have been accompanied by the monotonous beat of Etheridge's ram dismantling the last centre to be struck. At a meeting of the commissioners two days before Lovat's execution, Labelye announced that the centre had gone and Westminster Bridge stood entirely free.

> I have the hon. to report that in obedience to a former Order of this Hon. Board, on Saturday last the 4th inst. I caused the last Center of the Arches of Westmr. Bridge to be Eased and Struck which was perform'd with the usual dispatch & attended with the same success as all the other Arches. So that all the Arches of Westmr. Bridge . . . are now wholly supported by their adjoyning Piers and Abutments, & the whole Bridge stands free & entire in all its parts. I remain with respect the Rt Honble Comrs most Obedt Servt Chs Labelye
> Westm. 7th April 1747[18]

The surveyor's accounts record an award of £1 11s 6d given to the men on 4 April for striking the last centre. But it may be that though the centre was struck and the arch wholly self-supported, the timbers were not actually removed till July. This would have been unusual but Labelye, in *A Description* published four years later, says 'on 25th July 1747 the last Center was taken down and all the fifteen Arches of this Bridge were left entirely free and open'.[19] The last centre can be seen in a view by Canaletto painted for Sir Hugh Smithson, 'an excellent judge of the fine arts', at that time one of the commissioners and a regular attender at meetings, particularly between 1743 and 1745. Etheridge's centre is clearly shown with its ribs (only three of the five), puncheons and cross spars and a bucket dangling from the balustrade (Plate 15). The last four centres were on the Surrey side and were struck in September, November and December 1746 and April 1747.

Once the last centre had gone, the wooden fence across the Westminster abutment was taken down and re-erected across the middle of the bridge, the doorkeeper moving there at the same time.[20] All that remained was to finish the balustrade and recesses and to surface the carriage-way. 'It is now high time', Labelye informed the board in June, 'to provide a sufficient quantity of strong binding Gravel to make the upper coat of the Carriage Way of the Bridge and New Road and that the smallest quantity to be provided and furnished with all possible speed cannot be less than 500 loads. It is also high time to pave the Surrey abutment in the same manner as the West$^r$ abutment.'[21] A few weeks later orders were given to Jelfe and Tufnell and the paviour, Thomas Phillips, to pave the footway and pitch the coachway of Parliament Street and to pitch the Surrey abutment and north half of Bridge Street, Surrey, in the same way. The work was finished and Phillips's bill paid in August.[22] According to the *Westminster Journal,* 'the work on the South side of Westminster Bridge being finished there will be a passage over in a few Days'.[23]

A painting, also done for Sir Hugh Smithson, shows the bridge probably between April and June 1747 (Plate 35). The view is from the Stangate looking along the whole length of the bridge from the Surrey abutment arch to the Westminster abutment and causeway. The last centre has gone but work is still going on along the south balustrade which is complete as far as the east 56ft arch, stones being hoisted to the east 60ft balustrade.

By the autumn Westminster Bridge was finished down to the last of the twelve domed pedestrian shelters or recesses. There were recesses over each pier on both sides of the carriage-way but only twelve were covered. Labelye had produced drawings and a model in November 1746 and the commissioners had approved:

... the twelve of the Recesses over the points of the Piers of the Bridge, vid. the 4 center ones, 4 at the East end and 4 at the West end, be covered with Portland Stone, and that six of them be enclosed in front with a stone wall agreable to the Model this day produc'd by Mr Labelye.[24]

Jelfe and Tufnell's contract was signed the following February and witnessed by their clerk, Thomas Gayfere. The domes were to be ready in twelve months. They were to be semi-octagonal and to be fitted with consoles and keys at the tops to support the iron lampholders. The six enclosed recesses were to have circular windows above the doors with raised imposts to hold the carved sills.[25] All the recesses, not only the domed ones, were to be filled 'with Maidstone Blew Rock Rubble laid in mortar and pav'd with the best Purbeck paving three inches in thickness at the least with a sufficient drip towards the footway to carry off the Rain Water'.[26] Later on, when the bridge had been opened to the public, four oak benches were ordered for the four middle recesses, fixed with ribs on the tops and painted with four coats of oil. They cost five guineas each. Until the watch-houses were built near the abutments, the keys of the six enclosed recesses were kept by the beadles of St Margaret's, Westminster, and St Mary's, Lambeth, for use as temporary watch-boxes.[27] One of the enclosed recesses can be seen at the Lambeth end of Thomas Willson's engraving of 1747 (Plate 44) and in Malton's aquatint of 1790 (Plate 45).

On 14 November 1747 Labelye declared that 'the Bridge and the Roads and Streets on both Sides were compleatly finish'd, and the whole was performed in seven Years, nine Months, and sixteen Days from the laying of the first Stone'.[28]

But in May and June the first signs of calamity started to become plain. When the masons came to set the balustrade over the west 15ft pier (that is, the fifth pier from the Westminster abutment), it was found to be out of alignment. Something may have been wrong even earlier, during the winter when the *Westminster Journal* (but not the commissioners' Minutes) reported that the workmen were pulling down 'several of the Ballusters on the top of the new Bridge to set others in their room, they being too low'.[29] The balustrade was levelled and reset at the masons' expense but in a few days it was clear that the pier itself was moving. It settled gently at first but faster towards the end of July. No one at this point appeared to take much notice. There is no mention of anything amiss by the Works Committee, which in fact met only twice in June, twice in July and once in August. The commissioners themselves did not meet until December.

In fact, at a Works Committee meeting in June, Labelye was given a month's leave to visit Great Yarmouth. A petition had been presented to Parliament in February for leave to bring in a Bill to repair Yarmouth Harbour, and although it was passed in May and received the Royal Assent in June,[30] disputes had arisen. A letter from the Earl of Orford, Lord Townshend, Edward Walpole and Robert Townshend requested the commissioners' permission for Labelye to visit Yarmouth and advise on the condition of the harbours and piers.[31] The outcome of Labelye's journey was a report, *The Result of a View and Survey of Yarmouth Harbour taken in the Year 1747*, which was followed by another more effective Act of Parliament in 1750. But he was soon recalled to deal with the bridge. The situation was becoming serious. In the middle of August Richard Graham reported to Labelye a series of measurements he had been taking and mentioned that several piers had settled, in particular the west 15ft pier. He had noticed that open joints in the plinth and rail of the balustrade were increasing and 'that there was a visible bending in the Rail which was remark'd by most people & that the pavement had parted from the Plinth & opened'. Labelye was at that point going out of London again and told Graham to keep calm and that there was no danger. But the settlement continued until on 1 September it was 15in below the level of the two adjacent piers.

Graham's report, handed in to the Works Committee a week later, gives a sense of urgency and marks the beginning of the end of the generally friendly relationship which had existed between the Swiss engineer and his English staff.[32]

> The last time the Board met I told Mr Labelye, who was then going out of Town, of my observations, who bid me be easie and said there was no danger.
> As soon as I began to remark that there was an opening in the Purbeck above the Arch, I measured the Settlement which was then about 15 inches more than the Piers on each side, of which I wrote a letter to Mr Labelye, dated 1st instant, but he not being at the Gentleman's house to whom it was directed, it was returned back last Saturday. Expecting him every moment I delayed giving orders out of my province: but finding the Danger increase, insomuch that several of the stones out of the Soffite of the 64 & 68 ft Arches flaked off, and that the Pier settled very fast, and that part of a course of Key Stones was lower than those adjoyning about 5 inches, I thought further delays might be dangerous.

Graham decided that the time had come for action. He tried to get in touch with any commissioners then in London but they were all away. It was an extremely hot summer and no doubt London was thoroughly uncomfortable. However he consulted Jelfe, Tufnell and Etheridge, and ordered the fence to be moved back to the Westminster abutment and posted notices forbidding anyone to go on the bridge except those directly concerned with the works. Secondly he decided to ease the strain.

> I ordered that the Masons should take the Ballustrade, Cornice and Pavem$^t$ off the 68 & 64 ft Arches, and to lighten and ease the Pier as much as possible. I ordered that the Ballast should be taken off the Pier and the adjoining Spandrels, and obliged the men to work that night & on Sunday last. I likewise took care to keep Boats from going thro the said Arches, and desired Mr Holmes and Mr Etheridge not to part with the old iron and timber which I had sold to them, & I writ to all the Acting Commissioners near Town to beg the favour that they would be here today. The Pier is now about 20 inches lower than those on each side of it, but settles very slowly.
>
> R. Graham (Surveyor).

The Works Committee met on 8 September but even then only five members turned up: Colonel Herbert, Lord Stanhope, Kent, Laroche and Sir John Crosse.

By this time Labelye himself was worried and reported to the Works Committee at the same meeting that since the beginning of August the west 15 ft pier had begun to settle again,

> very gently at first but more sensibly all last week, the amount of all the former settlings being 7 inches and the present descent to this day being about 13 inches, but it has not settled sensibly since Saturday morning. I have every day since Saturday carefully examined that Pier and the adjoining Arches and find as follows—The Pier is perfectly fair and entire, level to a surprising degree, and in want of no repairs but raising it with new Portland to its due height when it has done settling. The two Arches resting thereon are both damaged but it is chiefly in the fronts, the inside Sophits of these two Arches having regularly followed the Pier that has settled.
>
> Upon the whole I am clearly of Opinion that the two Arches resting on that Pier are at present in no danger of falling and that they may be easily repaired without unbuilding them, provided the Pier does not settle much more.
>
> And that whatever may happen hereafter, the most prudent step to be taken at present is to order immediately a rough Center to be made and placed with all possible dispatch under each of these two Arches which will be ready and useful to all purposes, vid:
>
> If the Pier should settle much more it is not in the power of any mortal agent to hinder it or to hinder the Arches from following it, as long as it is

possible, and therefore in that case, the two Arches, instead of parting asunder and their materials falling in the River and not to be taken up without a great expense of time and money, they will be received and their material supported and saved in order to their being regularly rebuilt.

2nd If it is thought advisable at any time hereafter to unbuild and rebuild those two Arches, those rough Centers will be ready placed for that purpose.

And lastly if (as I hope) the Pier has nearly all the Weight it can have, it will not settle sensibly more, and therefore the Arches will only have their Zophiets and Fronts repair'd, those rough Centers will save almost all scaffolding which otherwise must be provided anew.

I hope this Board will not rise without issuing some general Orders for proceeding immediately on the Construction of those Centers and empowering their Officers for taking all other measures and precautions for preventing mischiefs and accidents and for supporting and maintaining the said Pier and Arches of the Bridge and for repairing all defects as soon as possible.[33]

Andrews Jelfe then gave the board his opinion that the arches were not in danger of collapse and that the pier should be loaded as much as possible to force it to sink as far as it could. He had found it to be leaning about 1in to the westward and believed that though the ground on which it rested must be faulty there was no spring underneath. The frame on which it was built seemed to be in good condition. He thought that the pier would not sink much more than 2in further but as a precaution the rough centres should be set up immediately, the pier loaded heavily, and the balustrade, plinths, cornice and pavement taken off both the arches.[34]

Samuel Tufnell gave similar advice except that he thought the superstructure should be taken off not necessarily to lighten the arches but to save material should the arches fall. He believed that both pier and arches could easily be repaired so that no mend need be visible, and that as the pier had sunk very little in the last two or three days it would probably not sink much more.[35]

Thomas Preston, the foreman to the masons, then delivered to the board the measurements he had taken morning and evening on the north and south fronts of the pier between 5 and 8 September, showing that the south front had settled $\frac{3}{4}$in and the north front $\frac{5}{8}$in. He did not think the arches would fall even if the pier should sink another 2ft, but a centre would certainly be a proper precaution.[36]

Robert Smith, the ballast man, was at a loss to account for the settlement but supposed there must be some quicksand underneath:

That when he dug the foundation of that Pier, he took all imaginable care; that he believed at that time it was a good Bottom; that by Boring it appeared to be so; and that he was so much satisfied with that foundation as with any of the others. That after the foundation was finished he Bored 6 or 7 ft deeper and apprehended what he Bored through was entire hard Gravel. That the Pier is above 25 yards from the Channel.

William Etheridge then gave his views, which were less optimistic than Labelye's or the masons'. He thought the settlement was caused partly by a quicksand along the western side of the river and partly because gravel for the roadway had been dug from too near the foundation. He was convinced that not only the two arches but the whole bridge itself was in danger and that he himself would not on any account venture underneath. Mr Graham had lightened the weight by removing a quantity of gravel from the spandrels and this was the only reason that the pier had been sinking more slowly. It would be a dangerous experiment, he thought, to load the pier at present even if centres should be erected under the two arches. [37]

The committee listened carefully to their officers' views. They resolved to endorse Graham's action and that the balustrade, plinth, cornice and pavement over the two arches resting on the pier and on the two ad-joining arches should be immediately taken down. The gravel between the spandrels should be removed so that the Purbeck counter-arch could be inspected. Labelye was ordered to do this at once and also to prepare a design with Etheridge for rough centres to be ready for the next meeting. Boats were moored on either side to prevent barges going through and a few days later these were replaced with floating booms fixed in position with piles. This precaution was just as well because early in the morning of 15 September, 'one of the great Stones, of several hundred weight, in the fifth arch, fell and made so great a noise into the Water that it affrighted the Watchmen. This stone has been loose and hanging out almost a foot for some time before'. [38]

At the next meeting there was an acrimonious discussion about the type of centre to be used. Etheridge, who at that time was hard at work on the design of Walton Bridge for Samuel Dicker, said that he could think of no better centre than that used before. They had been designed by himself and had answered their purpose perfectly, being a great deal more efficient than the centres originally designed by James King. But the two of them would take at least three months to build. Much to Etheridge's annoyance, Labelye then produced a design for centres which he claimed could be ready in six weeks. The Works Committee was impressed and ordered

Etheridge to build centres to Labelye's design for the two arches. The agreement was signed at the end of October, 'to build the two centres of Riga Timber in a good and workmanlike manner', and £500 was advanced on account.[39] The design of Labelye's centre can be seen in Canaletto's drawing of the bridge under repair in the Royal Collection (Plate 36), but Etheridge seems to have taken his time to build them. There was no time limit laid down in the contract. Labelye had told the Works Committee that they could be erected in six weeks, which meant they should have been ready well before Christmas, the contract having been signed on 27 October. But by March 1748 they were still only half made, Etheridge excusing his slowness because of the narrowness of the piece of ground near the abutment available to him for the work.[40] They were eventually finished in June.[41]

Meanwhile by the beginning of October the whole superstructure had been removed from the two arches except for the gravel in the spandrels, which Labelye recommended should be left till after the winter. During November the pier settled about another inch but still remained fairly level. A Portland header which had been loose for some time was still in place on 24 November and the soffits of the two arches seemed to be no more dangerous than they had been a fortnight before. As a precaution Labelye recommended that ballast should be filled into the pier excavations. He pointed out that the years when the piers and abutments were being built, 1739–43, had all been dry years and the river had silted considerably, bringing at least two feet of small sand into the excavations. On the other hand the years between 1743 and 1747 had been wet years and the Thames had scoured the sand away again, leaving the large ballast and gravel. Filling up again would merely be a prudent and proper precaution. During the fortnight before Christmas the rate of sinking seemed to be slower and there had been no more than ⅛in settlement.[42]

## NOTES TO CHAPTER 11

1. CJ 27 January and 20 February 1745/6, vol 25 pp 41 and 74.
2. Labelye *Description* pp 75–6.
3. Bridge Minutes 18 November, 23 December 1746.
4. *Gentleman's Magazine* XVI (1746) pp 683–4. Jeffrey's engraving was later used as frontispiece to the pamphlet *Gephyralogia* (1751).
5. Works Minutes 22 April 1746.
6. Reproduced in Constable *Canaletto* 752.
7. Bridge Minutes 19 March 1744/5 and 9 February 1747/8.

8. Ibid 29 April 1746.

9. *Westminster Journal* 11 October and 20 December 1746.

10. Bridge Minutes 8 April 1746.

11. *National Journal* 20–2 May 1746 (Bodleian Library).

12. Works Minutes 7 February and 22 October 1746.

13. Bridge Minutes 10 June 1746.

14. *Westminster Journal* 6 September and 29 November 1746.

15. Posted by order of the commissioners 16 December 1746 and published in most newspapers.

16. Bridge Minutes 16 December 1746 and printed in *General Advertiser* 17–19 December 1746.

17. CJ 5 and 9 February 1746/7, vol 25 pp 274 and 279.

18. Bridge Minutes 7 April 1747.

19. Labelye *Description* p 76.

20. Bridge Minutes 28 April 1747.

21. Works Minutes 2 June 1747.

22. Ibid 2 June, 21 July and 17 August 1747.

23. *Westminster Journal* 27 June 1747.

24. Bridge Minutes 25 November 1746.

25. Ibid 23 December 1746, 10 February 1746/7; Bridge Contracts (85) with Jelfe & Tufnell 21 February 1746/7.

26. Bridge Minutes 9 and 23 November 1743.

27. Works Minutes 20 November 1750; Bridge Minutes 30 April and 7 May 1751.

28. Labelye *Description* p 76.

29. *Westminster Journal* 27 June 1747.

30. Act 20 George II, c 40 (1747); CJ 4 February 1746/7 and a long discussion about Yarmouth Harbour 3 April 1747.

31. Works Minutes 16 June 1747. Lord Orford was the eldest son of Sir Robert Walpole, who had died of a stone two years before.

32. Works Minutes 8 September 1747, pp 200–1.

33. Ibid 8 September 1747 pp 194–5.

34. Ibid p 196.

35. Ibid pp 197–8.

36. Ibid p 198.

37. Ibid pp 198–9.

38. *Westminster Journal* 12, 19 and 26 September 1747.

39. Works Minutes 15 September and 27 October 1747.

40. Bridge Minutes 22 March 1747/8.

41. Works Minutes 7 June and 30 August 1748.

42. Bridge Minutes 17, 24 November and 15, 22 December 1747.

25. Westminster Bridge, between October and December 1745 by Antonio Jolli. *(Private Collection, London. Photograph by A. C. Cooper)*

26. Detail of Plate 25. Work on the arches at the Surrey end.

27. Detail of Plate 25. The bridge nearly finished.

# 12
# Westminster Bridge, 1748

The sinking pier afforded a topic of endless discussion from September 1747 onwards. Two questions occupied the minds of everyone interested in the subject, and that meant virtually every inhabitant of London and Westminster: what had caused the pier to sink, and how could a similar catastrophe be prevented from happening again?

It was not difficult to poke fun at the scandal and a flood of articles and pamphlets burst over the heads of the unlucky commissioners and their Swiss engineer. Thomas Touchit, editor of *The Westminster Journal,* in a long editorial found a comparison between Westminster Bridge and the national finances quite irresistible and made play with the Sinking Fund and the various 'piers' of salt, soap, candles, coffee, tea and chocolate on which Walpole's edifice of public credit was built. 'For I must observe', he says, 'that the Arches of this Political Bridge (the Aggregate, the South Sea and General Funds), like those of the *Stone Bridge* at *Westminster,* were originally intended to stand by themselves if there should be occasion: but by making them all one common structure according to the Great Man's Plan, the several parts have a mutual Dependence and Bearing on each other. Thus we see also in *Westminster Bridge* that the sinking of one Pier will prove the Ruin of two Arches.'[1]

Various popular ballads went the rounds and caused diversion to some and pain to others. The standard of versification was not high. 'The Downfall of Westminster Bridge or My Lord in the Suds' was a jibe at the Earl of Pembroke and contained such lines as:

> I sing not of Battles, not do I much chuse
> With too many vict'ries to burden my Muse;
> I sing not of Actions that rise to Renown
> But I sing of a monstrous huge Bridge tumbling down.[2]

Another ballad, sung to the tune of 'London Bridge is broken down', began

Westminster Bridge is sunk again,
  Dance o'er C-mm-ss-n-y;
The stones are almost rent in twain,
  C-mm-ss-n-y.
What must we do in this dreadful plight?
How set this sinking pier upright? etc.

and ended with the arrival of the ghost of James King who advised loading the pier with old cannon from Flanders and rebuilding it on piles:

And further, I think it would not be amiss
If, to add to their weight you threw on the Swiss . . .
If still it such monstrous weight beguiles,
As I told you before, you must build it on piles.[3]

King's name was also invoked in *Gephyralogia,* a pamphlet of 144 pages
published by C. Corbett in 1751 and sold for two shillings. *Gephyralogia*
began with a recitation of ancient bridges from the first bridge in Babylon,
Trajan's bridge over the Danube, and bridges in France, Italy, Spain and
Germany. It then listed about a dozen English bridges and described the
history of Westminster Bridge itself, using information taken from
Labelye's first two pamphlets and a supplement to the *Gentleman's
Magazine* for 1746. The anonymous author, on the whole sympathetic to
Labelye, pointed out that though Labelye had looked on piles under the
piers as a needless expense, yet there had been a strong body of opinion
among architects strenuously maintaining the contrary. James King he
mentions in particular, 'who not only urged the necessity of piling, but
offered to perform the work at such a reasonable price that disinterested
persons thought it not an instance of good economy to save such an
inconsiderable sum (about £5,000 or £6,000) in a work of such importance
and expense'.[4] Considering what eventually happened to the piers it
would have been 'common prudence to provide against it'. *Gephyralogia*
ended with an abstract of the rules of bridge building, citing Alberti and
Palladio; a short poem, 'When late the river-gods would visit Thames',
composed by 'a friend who deals in that way'; and an Appendix quoting
sections from the first three Bridge Acts. The frontispiece was 'A South
View of Westminster Bridge' by Thomas Jeffreys, Plate IX from the
*Gentleman's Magazine,* 1746.[5]

Batty Langley of course delighted in the disaster to his old enemy and
immediately published a shilling pamphlet, *A Survey of Westminster Bridge
as 'tis now sinking into Ruin,* in which he bludgeoned Mr Self-Sufficient

mercilessly for his 'unparall'd gross Ignorance, Madness and Knavery' in omitting to pile the foundations. He emphasised, with some justification, that if only the commissioners had taken notice of his own design, published in 1736, in which he had advised building the piers on 'substantial Piles of Oak, driven in as close as can possibly be done', the shocking spectacle of Westminster Bridge tumbling into mud and ruin need never have brought disgrace to the river Thames. He prefaced his protest with an engraved frontispiece (Plate 37) showing the bridge as it ought to have been built and enlivened with a scathing dialogue between the bridge master (perhaps Graham) and the Swiss engineer—'See here, thou Stupid Villian . . . D – m you for a Conceited Impudent Scoundrel, it would have been happy for the Publick had you been Hanged before West$^r$ Bridge was thought on. For by your Ignorance, Impudence & Villiany, the Publick is greatly Injured, & every Acting Commissioner is disgraced to the End of Time'. The luckless Mr Self-Sufficient is shown hanging from a gibbet beneath one of his own arches, labelled 'The Swiss Impostor rewarded as his Ignorance justly deserves'. Batty ended his pamphlet by declaring that the only possible course left open to the commissioners to save the other piers from a similar fate was to surround their foundation gratings with a close range of piles driven twelve or fifteen feet into the bed of the river. This, as it happened, was exactly what Labelye eventually did do, but Batty Langley's readers would probably have felt more sympathetic if his ideas had been expressed in less shrilly antagonistic language. No doubt the commissioners felt the same. As a reviewer in the *Gentleman's Magazine* put it, 'Mr Langley hurts his own cause by discovering too much resentment'.[6]

Even Dr Johnson, writing in *The Idler* ten years later, could not resist an uncharitable cut at the commissioners. Tom Tempest, one of the doctor's pessimistic stooges, who believed that nothing ill had happened by chance or mistake for the last forty years, thought that 'the arch of Westminster Bridge was so contrived as to sink on purpose that the nation might be put to charge'.[7] In its way it was a national disaster and many observers at the time were struck by the thought that their fine new bridge with its Portland stone gleaming in the autumn sunshine might perhaps be poised on the edge of ruin and about to plunge the entire nation into European ridicule. It is not surprising that, as Labelye said in his *Description,* 'many of the Commissioners returned to Town from their Country Seats on this melancholy Occasion' and asked the advice of as many people as they could, receiving schemes and proposals mostly 'absurd, improper or impracticable . . . but some far from despicable'.[8]

None of the originators of these schemes was mentioned by name in the Bridge Minutes. One can understand the commissioners ignoring the insults of Batty Langley, but others expressed their ideas in more tactful and restrained terms. William Halfpenny, for instance, produced a perspective view of the pier with a suggestion for relieving it of at least three quarters of its burden of stone and ballast. He recommended a support of four timber frames or trusses and six brick spandrel walls. Three brick arches were to spring from the walls and the voids were to be filled with pounded charcoal to preserve the wood, 'all which is humbly presumed will not only prevent the great weight of Ballast but throw a major part of the Burden on the sound Piers'. [9]

Charles Marquand, who had appeared a few years earlier with an unsuccessful machine for cutting piles under water, turned up again driving the guard-piles for keeping barges away from the sinking pier. He turned this practical experience to good use by publishing a pamphlet full of unhelpful suggestions for the future. His main point was that the stability of the foundations should have first been determined by driving a grooved pile into the soil to bring up specimens, 'for if you imagine the Stratum of Gravel to be deeper than it really is you run into an Error not easily remedied (and this I am afraid is the case of the failing Pier of Westminster Bridge) . . . What I have said', he added tactfully, 'is by no means intended as a Reflection on any Persons concerned in the building of Westminster Bridge. The Misfortune under which they at present labour might have befallen anyone'. He then advised the commissioners to enclose the foundation with a 'pallaplance' of dovetailed piles and concluded his pamphlet with a scheme to build the new piers at a convenient spot up or down stream, buoy them with barges and brewers' barrels, and float them to the bridge site. [10]

Marquand's ideas were instantly pounced on by an anonymous writer whom we have little difficulty in identifying as the indefatigable Batty Langley again, using his customary steam-roller technique to flatten the unfortunate 'Monsieur Pallaplance'. The idea of carefully boring the foundations with a grooved pile had been practised long before Monsieur Pallaplance came from Guernsey or Mr Self-Sufficient from Switzerland. And the method of enclosing the foundation with dovetailed piles had been described in detail, he says, in a pamphlet by one Batty Langley, *A Reply to Mr James* (1737), where 'on pages 26 and 27 Monsieur Pallaplance may see Rules long before laid down more particular and more eligible than his own'. The style of the author's picturesque jabs at Labelye confirms the hand of Batty Langley—'the never-to-be-forgotten Error, or want of Judgement in Mr Self-Sufficient', 'a glaring Instance of foreign Ability in the Ruins of Westminster Bridge', 'Mr Self-Sufficient the Foreign Engineer', and so on. Marquand's scheme for floating piers up or down stream he treats with ridicule. 'This weight is to be hung up with wood screws and hooks and eyes as Boys' Breeches are sometimes hung.'[11]

It may be that other proposals were put forward, possibly by the un-successful entrants of 1736, but they are not recorded in the minutes. Nor do the leading architects of the time appear to have been helpful in this crisis. William Kent died in April 1748, and Roger Morris, who had helped Lord Pembroke with the famous Palladian bridge at Wilton and had invested heavily in property near the Westminster approaches, died a year later. James Gibbs, also a considerable local property owner, 'had been long before much afflicted with the stone and gout' and was now in retirement. Men such as Isaac Ware, Henry Flitcroft and John Vardy were active and may possibly have produced suggestions or been consulted, although no record exists. John Gwynn, whose design for Blackfriars Bridge was adopted in 1759, was at this date known only as a writer on architecture.

This was a specialised job and the obstacles were daunting. Labelye's appointment specifically named him as engineer, not architect. But the method finally adopted for repairing the damage was put forward by an amateur, William Stukeley, and carried out with minor adjustments two years later. Stukeley was a clergyman and a distinguished antiquarian and field archaeologist known for his work on Stonehenge, Avebury and the Druids.[12] In 1747 he left his vicarage at Stamford, with its attractive garden, to become rector of St George the Martyr Bloomsbury. He im-mediately became interested in the disaster at Westminster and—no doubt with the help of his patron, the Duke of Montagu—was able to obtain access to the bridge and survey the damage. He produced a number of pen and wash drawings to illustrate his remedy of an internal relieving arch to take the weight off the pier, and of two quarter arches to relieve the thrust from the adjoining arches (Plate 38). In a later annotation Stukeley says that he handed in the drawings to the Royal Society on 27 October 1748 and before that to the Duke of Montagu, the Earl of Pembroke, Henry Pelham (the Prime Minister) and Labelye and Jelfe. He claimed that the design was accepted and 'put into execution'.[13]

Meanwhile the immediate problem was how to prevent the two arches from collapsing and the pier from settling any further into the mud. The usual petition was presented to Parliament at the beginning of the year and £20,000 was voted by the House of Commons to enable the commissioners to finish the work.[14] Early in January one of the keystones, which had been hanging loose from the soffit for some time, was taken out and lowered into a lighter and the remainder secured by a temporary wooden frame until the centres should be ready.[15] Work on the centres went ahead slowly, Etheridge still smarting under the affront to his professional skill by the rejection of his design in favour of Labelye's. Part of one centre capsized and lay bottom up in the river for some days. In February work began on fixing the centres in position but a severe frost caused ice to damage the guard-piles and booms, huge frozen blocks floating downstream like battering rams.[16] In March Etheridge reported progress and hoped to have the first centre set up in a fortnight. He said that everything was going well, 'especially if the damaged Arches should not be unkeyed; and even if the Arches were unkeyed he hoped the Centres would be able to support those Arches, tho they would receive a Blow thereby; but if the Arches should be unkeyed struts ought to be set up to strengthen the Centers and resist the Lateral Pressure'.[17] His account for material and workmanship from 15 September to 26 March for making and setting up the new centers came to £1,894 14s 2¾d.[18] In April workmen on the bridge were ordered to put in 'two hours in a Day extraordinary in order to hasten the finishing of it'.[19] The two centres were at last finished in June. The main cause of delay was the growing friction between Labelye and Etheridge. Unluckily the contract had been settled for daywork whereas formerly the centres had been contracted for within a specified time limit. 'The ill consequence of this', said Labelye, 'was that notwithstanding my most earnest Sollicitations and repeated Orders to use Dispatch, the two Centers cost great Sums of Money, and were above seven Months time in making and setting up; during all which Time nothing could be done towards repairing the damaged Pier and Arches'.[20]

In April 1748 Labelye, Jelfe and Tufnell reported to the Works Committee that they had considered what steps ought to be taken towards repairing the damaged arches as soon as the centres should be set up and had decided that the pier should be heavily loaded beyond the weight it was intended to bear.[21] A paragraph in the *Daily Advertiser* asserted that the pier was still sinking, whereupon the Works Committee examined Labelye, Tufnell, Etheridge and Graham, who all declared that the report was false and that 'they could not, by the best observations they could make, find the Pier had in the least sunk since January last'. This was confirmed by Preston, the masons' foreman, who said the pier had not moved for some time. From observations he had taken, he calculated the whole sinking amounted to 2ft 5in at the western end, and 2ft 2in at the eastern.[22]

The decision to load the pier was then put into force and 12,000 tons of old iron ordnance was requested from the Board of Ordnance and delivered according to an order from the Ordnance Office dated 17 May 1748. A few days later Labelye reported that the pier had been loaded with iron ordnance and rubble and that loading would continue till further notice. The two new centres were also finished except for a few braces. By 14 June the cannon had sunk the pier a further three inches and loading was postponed for a few days. By 21 June the mean level was 2ft 11in below the two adjoining piers. [23]

Unfortunately no painting or drawing is known of the cannon being loaded on the bridge. A satirical print by George Snow 'The X Plagues of Egypt', shows Labelye standing on a pier holding a cannon on each shoulder and exclaiming 'Morblieu'. [24] It must have been an extraordinary sight: 12,000 tons is a great deal of cannon even if only half of it was dumped on the bridge. Labelye said that 700 tons were actually used. Two views by Robert Griffier and Samuel Scott, both signed and dated 1748, show the bridge complete, undamaged and without ordnance. They were presumably painted before May when loading began, and tactfully ignored the sinking pier (Plates 39–40).

Tension was high and the undercurrent of hostility among some of the bridge officers came to the surface. Tufnell said he thought the centre would break if the pier was loaded any more. Labelye thought a further seventy-five tons of ordnance, then lying in the river on demurrage, could safely be loaded without endangering the centre. Etheridge agreed with Tufnell, reporting that the horizontal timbers were bent and 'the Center not strong enough but will be crush'd by the incumbent weight'. Graham also advised against further loading. Jelfe's opinion was not recorded but Graham complained that while carrying out the board's instructions to prevent people from crossing the bridge on Sundays he had been insulted by Jelfe, and after an investigation the committee decided that Graham's complaint was justified. [25] Labelye stuck to his opinion and obtained the support of four 'able and wholly disinterested carpenters' (William Ayres, Edward Gatton, Joseph Stibbs and James Neeld), who all agreed that the centres were in no danger. However the board decided to take no chances. They resolved that the centres should be strained no more and further loading should cease for a week for observation. [26]

A close watch was kept on the measurements and at a long meeting of the Bridge Commissioners on 5 July detailed reports, previously ordered by the committee, were delivered by Labelye himself, Jelfe and Tufnell, Graham, and the works carpenter, Etheridge. Labelye reported that he had bored the river-bed round the pier and had found the drill could be easily forced through the ballast reinforcement into a small kind of gravel which was less resistant to drilling than it had been in October 1739. Below this was a coarser gravel which was difficult to penetrate. In his opinion further loading would not be dangerous but he doubted if it would bring the pier to a firmer foundation. [27]

Jelfe and Tufnell's report was concerned mainly with the two arches. They thought that both the exterior Portland span and the interior Pur-beck span of the 68ft arch could support themselves without a centre, and in fact the centre then in position ought to be eased as soon as possible so that the pier could take its whole weight. The lesser arch of 64ft had been much disjointed by the settlement but this too was not in danger of falling and they recommended that its centre should also be eased. They did not think it proper for themselves as masons to pronounce on the safety of the centres, but 'as we know the Master Carpenters who have given it their joint opinion (that the Centers may be safely eas'd and lower'd) to be Men of Skill, great Experience in large Works, & quite disinterested in this Affair', they thought that loading the pier should continue even though the latest borings showed there was not much hope of finding a firm foundation.

Graham, the comptroller, handed in a longer and more detailed report, having 'no regard to anything but Truth and the safety of the Bridge'. He was firmly opposed to any more loading of ordnance and indeed thought the weight already on the pier should be taken off immediately because the centres were buckling, not only from the weight of the two neighbouring arches but also because of lateral pressure from the sides of the bridge. The centres were in great danger. To ease them would be extremely difficult because the blockings were twisted, the levels several inches lower at the east end than at the west, and the lower joints of the kingposts badly opened. But even if it were possible to ease the centres, he thought it dangerous to do so because it would inevitably mean the pier settling further into the river-bed and the collapse of the arches. 'I believe no one can answer for the Consequences', he said, 'either as to the Crushing the Center or the neighbouring Arches; neither can anyone, I believe, predict how soon those bad Consequences may happen.'

Etheridge's report ran along the same lines as Graham's. He thought the load should be taken off. To try to ease the centres would be both danger-ous and impracticable.

The commissioners were placed in a difficult position. The engineer, the masons, and a group of independent experts advised one course, the comptroller and the master carpenter advised another. They finally decided that the two arches should be dismantled and the pier taken down to water level, and that Labelye should immediately unload the cannon and send it back to the Ordnance Office. They also decided to unload the ballast from the spandrels, and that the two adjoining piers should 'be work'd up solid with rubble laid in mortar and be then loaded occasionally'. Having arrived at these decisions the commissioners adjourned until 6 December, leaving the Works Committee to cope with the situation for the rest of the summer and autumn.[28] Labelye maintained afterwards that the commissioners had been influenced by 'a wicked Cabal bent upon mischief' and that to unload the cannon was the only order he had ever received from the board 'contrary to my Judgement or Opinion and which I obey'd, but I own not without some concern'.[29]

Labelye himself was given three weeks' leave to visit Durham following a request for his expert opinion from William Whitfield, clerk to the Commissioners for Improving the Navigation of the River Wear.[30] The harbour at Sunderland had silted up and a petition presented to the House of Commons in January 1747 was followed by an Act designed to relieve the 'great obstruction and Discouragement to the Trade and Navigation' it caused.[31]

Labelye returned from the north to inform the Works Committee that he himself, the masons and Etheridge were all agreed that the Purbeck or secondary arches should be pulled down 'and that no mischief could follow therefrom'. The committee agreed and ordered him to begin work immediately and also to lay before the board in writing his ideas for repairing the damage in 'the most speedy and effectual manner'.[32]

Labelye's report covers fifteen closely written folio pages and was handed in at the commissioners' first meeting of the winter on 6 December 1748. It was accompanied by drawings and models illustrating his ideas on how the damage could best be put right.[33] His proposals crystallised into two schemes, the first of which was adopted a month later; the second was scarcely even considered, and had been put forward, one suspects, with the psychological intention of persuading the commissioners to accept the first scheme. The report begins with the recital of half a dozen harebrained suggestions which Labelye did not even trouble to demolish. It then explained his belief that the commissioners must first decide whether the pier could be wholly trusted as it was or whether it should be strengthened and the foundations secured.

His first scheme, and that finally agreed on, involved the pier being strengthened by means of three rows of piles driven round it into the river-bed in order to compress the foundation. The pier and springers could then be rebuilt to the upright and in range with the others. Two centres should be set up, the distorted centres built by Etheridge earlier in the year being used again with modifications. The Portland arches could then be rebuilt and as soon as they were keyed in, the centres should be eased. On these two Portland arches, secondary Purbeck arches should be bonded in as before but with the difference that they should be equal throughout the arc instead of being thicker at the reins than at the key. They should be joined and reinforced by a reversed arch over the pier. The spandrels were not be be filled in with ballast as before but a scheme arch and two half arches should be built over the pier, about ten feet on each side and abutting on the two adjacent piers. This would lessen the weight, which could be further reduced by rebuilding the octagon turrets slightly less thickly and without rubble filling, and by not rebuilding the rib walls over the arches. The superstructure could then be set up again using the same materials. A sectional drawing of all this, by Thomas Gayfere, is used as an illustration in a copy of Labelye's *Short Account*, formerly belonging to Rennie and presented by his widow to the Institution of Civil Engineers.

Labelye maintained that this scheme was both practicable and sensible and with the advantage that it preserved the original symmetry of the bridge. To the hypothetical argument that symmetry was insignificant compared to safety and that there was really very little guarantee that the pier might not settle even further, Labelye replied that the loading and piling would compress the ground to such an extent that the foundation would be 'confin'd and greatly secur'd from any danger of the Water acting on it at any time hereafter'.

The second scheme involved building two piers, one on either side of the settled pier, and putting up three pointed arches. The superstructure would be the same and the stone from the old arches could be used again. New stone would have to be found for the new piers. Labelye felt it his duty to tell the board the various objections which might be offered against this scheme, together with his answers, but his heart was so patently not in its favour that all this section of the report rings hollow. 'I conclude with declaring', he said, 'that in my opinion this Scheme (the first) is both proper and practicable and that I think it would prove entirely effectual.' He then came as near as he ever did to an admission that he had made a mistake in the basic design of Westminster Bridge. 'The nature of the ground in that place has prov'd defective', he said, 'tho it is not exactly known, and perhaps never will be known, wherein it is faulty.'[34]

Jelfe and Tufnell were called in to the meeting, and told the board that they had seen both schemes and were in complete agreement with Labelye, but the commissioners were not wholly convinced. They ordered both Labelye and Etheridge to report their opinions on what might happen to the two adjacent arches (72ft and 6oft) should the damaged arches be dismantled. Etheridge reported a week later that he thought the arches would certainly be in danger from lateral pressure but that he proposed to counter this by shoring them under the crowns. He also suggested a modified form of Labelye's first scheme, making use of a caisson to repair the pier. The Board then resolved that Labelye's first scheme should be carried out, 'to the great Disappointment & Mortification of the whole Cabal, and of a great many other Projectors'.[35]

Labelye handed in his own report at the next meeting, just before Christmas, firmly stating that the adjacent arches were in no danger whatever. All arches, he said, have a tendency to lateral pressure or thrust, but

> the Arches of Westminster Bridge have been design'd by me & have been built upon very different principles; they are each of them double, the undermost a semi-circular arch of large blocks of Portland Stone. Now all persons in any degree versed in the practical parts of Building know that even semi-circular arches have some lateral pressure (though the great Palladio thought otherwise) but that it is inconsiderable & that a very small load on the piers, less than the tenth part of what the piers of Westminster Bridge are now loaded with is sufficient to destroy or overbalance it.

His system of secondary Purbeck arches, he maintained, brought all the arches to a state of equilibrium and therefore not only were the 72ft and 6oft arches in no danger, but Etheridge's idea of shoring them up with new centres would be expensive and totally unnecessary. The board decided to consider all this in the New Year, and the clerk was ordered to summon the commissioners by advertising in the newspapers.[36]

They met again on 10 January 1749 and, considering the importance of the decisions to be taken, the number of commissioners attending was regrettably small. Lord Pembroke took the chair assisted by the Duke of Richmond, the two Horace Walpoles, Lord Charles Cavendish, Edward Smith, Sir Hugh Smithson, Sir Thomas Hales, William Glanville and Welbore Ellis. They negatived a motion that consideration of Labelye's report should be postponed till a more representative meeting could be held, and resolved, seven for and one against, to accept the report.[37]

## NOTES TO CHAPTER 12

1. 'Humorous Reflections' in *Westminster Journal* 12 September 1747, and reprinted as 'A Lucubration on the Sinking of Westminster Bridge' in the *Gentleman's Magazine* 1747 p 533.

2. 'The Downfall of Westminster Bridge or My Lord in the Suds, A New Ballad to the Tune of King John and the Abbot of Canterbury, printed for H. Carpenter in Fleet Street (Price Six-pence)', undated but about 1747.

3. *London Magazine* August 1748 p 374.

4. James King's generous offer is not mentioned in the Bridge Minutes.

5. *Gephyralogia, an historical account of Bridges antient and modern: including a particular history and description of the new Bridge at Westminster* ... (1751).

6. B. Langley *A Survey of Westminster Bridge as 'tis now Sinking into Ruin* ... (1748); *Gentleman's Magazine* 1748 p 96.

7. Samuel Johnson *The Idler* No 10 (17 June 1758).

8. *London Magazine* (1747) p 434; Labelye *Description* pp 77–8.

9. William Halfpenny *A Perspective View of the Sunk Pier and the two Adjoyning Arches at Westminster* ..., engraved by Parr and published by Brindley 1748. Halfpenny was a builder with a turn for writing. He had published a number of guides and builders' handbooks under the pseudonym of 'Michael Hoare'.

10. Charles Marquand *Remarks on the Different Construction of Bridges and Improvements to secure their Foundations* (1749).

11. Anonymous but probably Batty Langley *Observations on a Pamphlet entitled 'Remarks on the Different Constructions. . .' in which the Puerility of that Performance is considered* (1749).

12. Stewart Piggott *William Stukeley* (1950).

13. Stukeley's drawings in Bodleian MSS Top Gen b53 fol 90 and Gough Maps 23 fol 57b; British Museum Crace Collection (Views) V 93.

14. Bridge Minutes 27 January 1747/8; CJ 15 February and 4 April 1748, vol 25 p 614.

15. *Westminster Journal* 9 January 1747/8.

16. Ibid 30 January, 13 and 20 February 1747/8.

17. Bridge Minutes 22 March 1747/8.

18. Ibid 29 March 1747/8.

19. *Westminster Journal* 9 April 1748.

20. Labelye *Description* pp 78–9.

21. Works Minutes 12 April 1748. Loading the pier had been a subject of dispute since the previous autumn (*London Magazine* (1747) p 529)

22. Works Minutes 10 May 1748.

23. Ibid 7, 14 and 21 June 1748.

24. *British Museum Political and Personal Satires* 3020.

25. Works Minutes 12 July 1748.

26. Bridge Minutes 28 June 1748.

27. Ibid 5 July 1748.

28. Ibid 5 July 1748.

29. Labelye *Description* pp 80–1.

30. Works Minutes 19 July 1748.
31. Act 20 George II, c 18 (1747). Labelye's visit resulted in his *Report relating to the Improvement of the River Wear and Port of Sunderland* (1748).
32. Works Minutes 27 September and 29 November 1748.
33. Bridge Minutes 6 December 1748. Unfortunately none of Labelye's many drawings and models used during the building of Westminster Bridge accompany the Bridge Records in the PRO, nor in the copies in the House of Lords.
34. Bridge Minutes 6 December 1748.
35. Ibid 13 December 1748; Labelye *Description* p 82.
36. Ibid 20 December 1748.
37. Ibid 7 January 1748/9.

# 13
# Westminster Bridge, 1749

In October 1748 the War of the Austrian Succession came to an end with the Peace of Aix-la-Chapelle—*bête comme la Paix* as the wits of Paris said—and Europe entered into eight years of uneasy peace during which England under the Pelhams and Pitt prepared for the age of imperial expansion. For Westminster Bridge the peace came too late. The transport of stone up-Channel from Dorset had for years been subjected to hazard and delay waiting for convoys; but by 1748 the full quantity of stone needed for the works had been delivered and there was enough lying in the wharfs to cope with repairs to the damaged pier and arches. Watermen on the river, especially down by the Tower and below London Bridge, had suffered severely from the attentions of the press-gangs, though in general men working on the bridge had been immune. William Wotten, one of Robert Smith's ballast men, was impressed, however, and his father successfully petitioned the commissioners for his release:

> The Humble Petition of Thomas Wotten most humbly sheweth that your Petitioner's son, William Wotten was Imprest into his Maties Service soon after I contracted with yr Hons for raising Balast, and since cast away on board the 'Portsmouth' and is now on board his Maties Ship the 'Culloden', now lying at the Nore & being no Sailor is now in a deplorable condition.

The commissioners ordered Ayloffe to write to Thomas Corbet, Secretary to the Admiralty, demanding Wotten's discharge.[1] The peace threw dozens of ships out of commission and hundreds of watermen returned to the Thames to add to an overcrowded profession and increase a potential source of danger to the bridge works.

Peace had been signed in October 1748 and preparations for the celebrations lasted throughout the winter. It was proclaimed by the heralds in February and the actual celebrations took place in April. There were illuminations, 'a jubilee masquerade in the Venetian manner' at Ranelagh, and an ambitious firework display in Green Park laid on by Charles

Frederick, Comptroller of the Fireworks, and the Chevalier Servandoni, who had designed an elegant pavilion of wood and whitewashed canvas decorated with pictures, statues, a balustrade, festoons of flowers and other delights.[2] The fireworks themselves, eagerly awaited by coachloads of visitors pouring into London for the celebrations, were an anticlimax spoilt by slowness, not enough change of colour and shapes, a fire which devastated one whole wing of the pavilion, and a quarrel between Frederick and the Chevalier in front of the royal party. The Chevalier had to be disarmed and made to apologise before the Duke of Cumberland next morning. Desaguliers' son, Captain Thomas Desaguliers, turned up in the capacity of chief fire-master of His Majesty's royal laboratory.[3]

There was a great deal of discussion about the propriety of celebrating so flimsy a peace with such ostentation;[4] but there was no doubt at all about the success of an event which took place a few days earlier. This was the rehearsal in Vauxhall Gardens of Handel's 'Music for the Royal Fireworks', commissioned by the king to commemorate the peace. Handel had already composed a Te Deum in honour of George II's personal triumph at Dettingen, and the king sensibly did not intend to lose another opportunity. Handel was set to work, and—in exchange for providing lanterns and thirty skilled operators to help with the Green Park fireworks—the owner of Vauxhall Gardens, Jonathan Tyers, was allowed to hold the public rehearsal. This was a tremendous success and some 12,000 people assembled at Vauxhall on 21 April to hear this martial masterpiece.[5] Most of the traffic went round by London Bridge, which became so congested that carriages were held up for three hours. The watermen of Westminster and Chelsea made full use of their stroke of good luck, landing hundreds of revellers at Vauxhall Stairs. In fact, Westminster Bridge being temporarily out of action came as a boon to discharged seamen returning to the Thames from ships lately put out of commission, and the price of boats went up by at least a quarter.

In spite of its dangerous condition scores of people attempted to cross the bridge and several accidents happened when the workmen tried to carry out the commissioners' orders to keep intruders away. Warning notices had to be published in the papers and stuck up on the abutments:

> Ordered That no person be permitted to pass the Bridge on any pretence whatever and that the Doorkeepers do not suffer any Person to come upon the Bridge except the proper Officers, Artificers and their workmen, and all the Commissioners are desired not to go upon the Bridge without taking with them an Officer belonging to the Commission, it being otherwise impossible for the Doorkeepers to distinguish the Com.rs from other Gent.[6]

Meanwhile another £12,000 was voted for the bridge by Parliament. Lord Trentham, one of the Members for Westminster, presented the petition on behalf of the commissioners and confessed to the Commons about the mishap to the pier. In spite of the utmost care, he told them, 'just at the time when the Bridge was nearly finished and made passable, one of the Piers thereof, by some unforseen and unknown Accident, sunk so considerably that the Petitioners find themselves under an indispensable Necessity to repair the Damage sustained by the sinking thereof'.[7]

In January the board had agreed to accept Labelye's report but it was not until March that he was ordered to unkey and take down the Portland arches and go ahead with his plans for rebuilding:

> Order'd That Mr Labelye forthwith proceed to unkey and take down the Western Portland 60 & 64 ft Arches of the Bridge and lay the materials thereof on the several wharfs provided for that purpose, and carry into execution the several Resolutions of this Commission of the 5th July and 13th December last.[8]

Two contracts had already been settled. John Smith of Lambeth had leased a wharf and two cranes near the Stangate at £150 a year from 10 March; and Samuel Price, the ballast heaver who had worked on the bridge since the foundations were dug in 1738, agreed to move the stone to the wharfs at one shilling a ton, finding boats, men and all other necessaries himself.[9]

Jelfe and Tufnell's contract for repairing the damage appeared as a draft for the Works Committee in June but was not actually signed until the following February. Provided the stone was brought from the wharf and they were allowed to use equipment belonging to the commissioners ('windlasses, tackelfalls, blocks, handscrews, crowes . . .'), they agreed to rebuild the arches with a cornice and balustrade at the usual rates 'within the space of twelve Kalendar Months next after the day on which the several wooden Centers and hoisting scaffolds shall be set up'.[10]

At the end of May Labelye informed the board that the masons 'had taken down all Portland Stone of the damaged Arches since 20th May so that in a few days they must dismiss their men excepting some few which may be employd in converting the broken Stones into useful ones'.[11] A drawing by Canaletto at Windsor Castle shows the work at this stage, viewed from the end of the Westminster south causeway looking along the length of the bridge towards Lambeth (Plate 36).[12] Labelye's, not Etheridge's, centres are shown under the two damaged arches, which are

28. *View of London in 1746 by Canaletto. (National Gallery, Prague. Photograph by courtesy of the National Gallery, Prague)*

29. Detail of Plate 28. Work on the bridge between June and November 1746.

30. Detail of Plate 28. Centres at the Surrey abutment.

both being dismantled. The 15 ft pier is considerably below the level of the others. The secondary Purbeck arches had already been taken down the previous October but the Portland arches do not appear to have been unkeyed, so work had not progressed far since 14 March, when Labelye was given leave to begin demolition. The drawing was probably done in April and certainly some time before 20 May, when Labelye reported that the Portland arches were down.

Two important changes in the staff occurred at this point. Richard Graham, who had served as comptroller and surveyor since 1738, died on 8 May; and William Etheridge, master carpenter of the bridge since 1743 and James King's foreman before that, was dismissed. Shortly before his death Graham's salary had been reduced from £300 to £290, possibly because of illness and inability to work full-time. His successor, Obadiah Wylde of Ratcliffe Cross, was appointed Clerk of the Cheque in addition to the post of comptroller and surveyor, and had strict orders to keep daily accounts of time and work done on the bridge. Labelye told the committee that Wylde seemed 'to be a very skilful person and did not want any other instructions than such as he has already received from the board'.[13] Edward Ruby, the new carpenter succeeding Etheridge, was 'a water Carpenter who had been several years employd in the Works of Dagenham Reach and Rye Harbour'.[14] Etheridge had handed over the keys some time before this and the change came as a relief to Labelye, who had probably expected more loyalty from his staff. Etheridge and Graham seem to have formed the nucleus of a group described by Labelye as 'a wicked Cabal bent upon Mischief' and determined to hinder the works as much as possible. It had credit enough, says Labelye, to make some of the commissioners believe that loading the pier with cannon would be dangerous to the arches and crush the centres. 'About the end of April 1749 by the removal of one person and the death of another, which happened soon after, the Cabal was totally routed, its mischievous Intentions seen through and prevented, and Peace and Harmony returned again.'[15]

As soon as the masonry was down and out of the way on the wharfs, work began on driving the piles to consolidate the foundations. The comptroller was told to find out about the current prices of timber and to choose '200 pieces of the best straight Riga timber about 12 inches square at the small end, none less than 28 ft long and none more than 32 ft.' It was supplied by Leonard Phillips for £76 7s 6d on 20 April.[16] Vauloué's pile-driving engine was brought out of storage and the contract was renewed with Robert Halliwell to provide three horses and an able driver for three months on the same terms as before.[17]

Work began on 17 July but the next day some sort of accident happened to the engine and it was early in August before it was ready again. In November Labelye was able to report that the dovetailed piles had been driven with such dispatch that only eighteen or nineteen were left. By working at every tide almost day and night the whole were driven and sawn off underwater by 20 December, the commissioners recording their appreciation in the Minutes:

> Mr Labelye Engineer and Mr Rubie Carpenter reported verbally that the Dove-tailed piling round the foundation of the settled Pier was compleated this Day.
> Resolvd It is the opinion of this Board that their Officers and Mr Rubie the Undertaker have performd their Duty in this work with great Application and Judgement.

Labelye also reported that the ground had been so compressed that in his opinion the other two rows of piles were unnecessary.[18]

The year 1749 ended with more election troubles in Westminster. Viscount Trentham, who had been a Member for Westminster since 1747, was appointed one of the Lords Commissioners of the Admiralty and had to stand for re-election. Trentham had been unpopular in some quarters. He was young and modish, could speak French, and had befriended a troupe of French players in trouble at the Haymarket Theatre. This, and the fact that he was Earl Gower's son, enabled his opponents to accuse him of Jacobitism and disloyalty to the Crown and gave rise to a string of satirical ballads, such as 'Peg Trim Tram in the Suds: or no French Strollers', and a number of prints insinuating bribery, such as 'Britannia Disturb'd or an Invasion of French Vagrants, address'd to the Worthy Electors of the City of Westminster'. At the close of polling on 8 December Trentham was declared elected but the defeated candidate, Sir George Vandeput, demanded a scrutiny. He protested at the conduct of the election (the hiring of ruffians disguised as sailors, the vote being allowed to paupers, and so on) resulting in the High Bailiff registering a 'false and Pretended Majority in favour of Lord Trentham'. He petitioned the House of Commons and the quarrel dragged on until April 1751 when after a long debate two of the ringleaders, Alexander Murray and John Gibson, were committed to Newgate for inciting the mob on behalf of Vandeput. Murray was also penalised for refusing to kneel at the Bar of the House to receive the Speaker's reprimand.[19]

## NOTES TO CHAPTER 13

1. Bridge Minutes 15 December 1747.
2. The 'Machine for the Fireworks' is illustrated in colour in Alan St H. Brock *A History of Fireworks* (1949) p 52.
3. The fireworks fiasco is described by Horace Walpole to Horace Mann 3 May 1749. *Gentleman's Magazine* (1749) pp 185–7 and *London Magazine* (1749) pp 192–4.
4. A print appearing on 27 April, 'NO Money with Fireworks, Money with Commerce', shows an Englishman biting his nails in vexation, his pockets empty and turned inside out, and a smug Dutchman with his pockets overflowing (*British Museum Political and Personal Satires* 3028).
5. For the correspondence between the Duke of Montagu, Master of the Ordnance, and Charles Frederick, Comptroller of the Fireworks, on the subject of Handel's objections to the Vauxhall rehearsal, see Otto Deutsch *Handel, a Documentary Biography* (1955) pp 661–9.
6. Bridge Minutes 26 April 1749; *Westminster Journal, General Advertiser* and other papers.
7. CJ 9 February and 12 April 1749, vol 25 pp 731, 834.
8. Bridge Minutes 14 March 1748/9.
9. Ibid. Smith's contract was extended for another year to 2 March 1750.
10. Works Minutes 13 June 1749; Bridge Contracts (88) with Jelfe & Tufnell 6 February 1748/9.
11. Works Minutes 30 May 1749.
12. Reproduced in Constable *Canaletto* 751.
13. Bridge Minutes 21 March 1748/9.
14. Ibid 26 April 1749.
15. Labelye *Description* pp 80 and 83.
16. Works Minutes 4 April and 23 June 1749.
17. Bridge Contracts (87) with Halliwell 1 July 1749.
18. Bridge Minutes 5 December 1749. Labelye *Description* p 84
19. CJ 18 January 1749/50 and 1, 12 February to 2 April 1751; *Parl. History* 14 (1747–53) pp 761 and 870–901; *British Museum Catalogue of Political and Personal Satires* 3043–4.

# 14
# Westminster Bridge, 1750

Early in the New Year the bridge staff were saddened by the death of the Earl of Pembroke, which happened shortly after a meeting in the Bridge Office on 9 January. Horace Walpole (senior) had taken the chair and thirteen other members were present.[1] They had approved an adjustment to Jelfe and Tufnell's accounts, a conveyance of some property in Bridge Street, warrants for salaries for Ayloffe, Lediard, Bowack, Seddon and Lewis, and had discussed at some length the exact position for the new Fishmarket, deciding that Lediard and Simpson should meet the Fishmarket Trustees and stake out the boundaries.[2] It was an average meeting lasting for about two hours but the effort had proved too much and Pembroke died that evening in his house in the Privy Gardens. Horace Walpole wrote to his faithful correspondent, Sir Horace Mann:

> Lord Pembroke died last night; he had been at the Bridge Committee in the morning, where according to custom, he fell into an outrageous passion; as my Lord Chesterfield told him, ever since the pier sunk he has constantly been *damming* and *sinking*. The watermen say today that now the great pier (peer) is quite gone.[3]

Pembroke's death was a blow to Labelye who for many years had relied on his powerful interest for support in time of crisis. He recorded his decease 'not long ago to my very great Sorrow'. An analysis of the meetings shows that from the first he had been in constant attendance and had been chairman of commissioners' meeting no less that 120 times, and of the Works Committee meetings thirty-six times. Most of his contemporaries interested in Westminster Bridge seem to have agreed that not only the day-to-day affairs of the bridge benefited from his practical experience of architecture; the whole concept of the bridge itself, its difficult birth in the face of powerful hostility, its prosperity during the long years of war with France, its design and even its decorative detail, were in great measure due to the Earl of Pembroke. Roubiliac's marble bust of him at Wilton (Plate 16) is inscribed:

Qualis ille vir fuerit Si caetera faceant
Loquantur aedes Wilton^ae Et pons ille
Westmonast^s 1750 Hen: Com: Pembr: & Montgom.

There is no record of who replaced him on the commission, though there is a note that Thomas Walker, Surveyor of the Crown Lands, who also died at about this time, was succeeded by Sir Peter Thompson of Bermondsey, MP for St Albans.[4] Pembroke died suddenly, and curiously enough some verses—no doubt intended to celebrate his laying the first stone almost eleven years before—were written on the very morning of his death, before the news got about.

Who e'er this mighty frame surveys,
Must join in Pembroke's ceaseless praise,
His steady care his active heart
Produc'd this noblest work of art.
The fair approach to him we owe,
Oppos'd by every wile of law;
Vexatious claims he caus'd to cease
And legal feuds to end in peace . . .

The lines were dedicated 'to the Rt Hon the Earl of Pembroke, the noble Patron and Director of the Bridge at Westminster.'[5]

Meanwhile, according to a resolution of the commissioners, Admiral Warren presented the customary petition to Parliament; the Chancellor of the Exchequer told the House that the king had recommended it for their consideration, and the sum of £8,000 was voted in Supply to carry on the work.[6] At the same time the commissioners were compelled to present a memorial to the Treasury 'desiring their Lordships would be pleased to direct that the sum of £12,000 granted the last Session of Parliament towards building the Bridge may with all convenient speed be paid into the hands of Mr Sam. Seddon, their Treasurer, without deducting the usual Fees payable to the Civil List'.[7] Seddon's weekly accounts for March 1750 show that he did receive the £12,000 but that £3 10s 6d was deducted for fees paid at the Treasury and Exchequer.[8]

During the early part of the year Ruby's men were hard at work altering the centres and in February Labelye told the board that they were nearly ready. The board ordered that 'the said Centers be set up with all convenient speed and that Mr Labelye do from time to time issue the proper orders for that purpose'. A few days later one of the feet of the first centre was fixed in the river.[9]

The bridge was subjected to another hazard in the spring of 1750. On 8 February two unusually severe earthquake shocks were felt in England, particularly in London and Westminster along the banks of the Thames. People in the courts in Westminster Hall fully expected the roof to fall, and the new buildings round Grosvenor Square shook so much that the occupants ran terrified into the streets. Another shock was felt exactly a month later, on 8 March, 'attended with a great rustling noise as of wind'.[10] Fortunately the bridge stood firm and the shocks made no impression even on the damaged arches. Labelye was able to announce that they had been quite unaffected, 'to the great amazement of many and the no less Confusion and Disappointment of not a few malicious and ignorant People who had confidently asserted and propagated the notion that upon unkeying any one of the Arches the whole Bridge would fall'.[11] A third shock was gloomily expected on 8 April, based on the prophecies of Mitchell, a mad guardsman in Bethlehem Hospital. Hundreds of gullible Londoners flocked into the country to escape the horrors of being swallowed alive. A satirical print on the subject was issued, 'The Military Prophet: or a Flight from Providence. Address'd to the Foolish and Guilty, who *timidly* withdrew themselves on the *Alarm* of another Earthquake April 1750'.[12] And Horace Walpole ridiculed 'the frantic terror prevailing so much, that within three days seven hundred and thirty coaches have been counted passing Hyde Park Corner, with whole parties removing into the country'.[13]

Shortly after the earthquake scare an outbreak of timber thieving occurred in the wharfs, and Labelye reported the loss of three short puncheons of fir timber, part of the last rib of one of the centres. 'As these pieces were worth little or nothing and were better at hand, it seems to be done wilfully or maliciously by a person well acquainted with the Yard and how the taking these away then would occasion the most delay in the putting up of that Center. Either the Watchman did not attend or his Dog was not loose or else the Thieves came in with a Boat at high water'. An advertisement offering a twenty-guinea reward was posted in the *Daily Advertiser,* but a week or two later, Etheridge (no longer employed on the bridge but still living in Westminster) said that he had seen the journeymen carpenters carry off loads of timber from the works. This was hotly denied by Ruby and Wylde, both of whom declared that Etheridge's allegation was 'entirely false and that great care was taken to prevent any Waste or Embezzlement being made of Timbers belonging to the Commission'.[14] The thieving seems to have been a minor affair and no more was heard of it. By 18 April the new centres and hoisting stage were in position and Jelfe and Tufnell were able to begin rebuilding the arches. They had been working for months on the pier, restoring it to the perpendicular. This was done by chipping with hammer and chisel and was a long arduous business not made any easier by the masons having to work at water level in January, February and March. But with this finished and the centres ready they were able to start work on the arches. They began on 23 April, and though the contract allowed them twelve months, by means of employing at least 100 men both arches were keyed by 20 July. The centres were struck a week later and 'neither of the new Arches descended or followed their Centers sensibly, not even so much as the Thickness of a common Packthread'. [15]

At a meeting in June the gatekeepers were ordered to keep intruders off the bridge, which by this time was once again so nearly finished that it offered a tempting alternative to a passage across the river in a boat rowed by an often foul-mouthed and rapacious waterman:

> Ordered That the Gate Keepers upon the bridge do not under any pretence whatever suffer any Person to pass over or come upon the Bridge, the Officers and Workmen employed in the Service of this Commission only excepted; and in case either of the Gate Keepers shall offend against the above Order the said Officers and Workmen or either of them acqu. Mr Seddon the Treas. therewith who is hereby directed immediately to discharge the Gate Keeper so offending and to appoint another in his stead.

The order appeared fortnightly in the *Daily Advertiser* and was also fixed on the abutments. [16] There was no further meeting of the commissioners until October, though the Works Committee met twice more during the summer, when Labelye was ordered to strike the centres of the counter-arches. [17]

In October Labelye reported that the masons had completed most of the work on the arches 'in a good and workmanlike manner and entirely to my satisfaction and that the amount thereof is near £7,000 of which they have already received £5,000'. The Treasurer was authorised to pay them £1,000 on account and the remainder was settled on 13 November. In fact the repairs were so nearly finished that the commissioners called a meeting to decide when the bridge should be opened officially. The officers and foremen were asked to advise, and assured the board that the work should be completed in a week or ten days. 'A Motion being made and the Question put it was unanimously Resolved That the Bridge should be opened for all Passengers Horses and Carriages on Saturday the Eighteenth Instant.' The resolution was signed by the clerk and published in the *Gazette* and other newspapers.[18] The last stone was laid on 10 November by Thomas Lediard in the presence of a group of commissioners.[19] Four men were employed to fill in the ruts every day and to prevent people damaging the superstructure, which, in spite of the fearsome penalty threatened in the Act, seems to have been as popular a pastime then as it is now. Finally the fences were taken away from each end and Labelye ordered to remove 'all other obstacles and make the Bridge compleatly passable by Sunday next'.[20]

The opening was widely advertised in the papers and in spite of a few protests expressing disapproval of Sunday holidays, the commissioners stuck to their decision and arranged for the ceremony to begin at midnight so as not to deprive the Londoners of a moment's merry-making. A reporter on the *Whitehall Evening Post* described the scene graphically:

> On Saturday last a great Number of the Principal Inhabitants of the Parishes of St Margaret's and St John's Westminster met at the Bear at the Bridge foot, where having dined the following Toasts were drank, viz. the King with a discharge of 41 Pieces of Cannon, accompanied with a flourish of Trumpets and Kettle Drums; the Prince and Princess of Wales with 31 ditto; the Duke and the rest of the Royal Family with 21; the Pious Memory of Queen Elizabeth with 41; the Hon Members of the City and Liberty of Westminster, and the rest of the Commissioners of the Bridge, with 31; And having spent the Evening with much Mirth and merry Songs, particularly two new occasional ones, with repeated Huzzas, at half an Hour after 12 they march'd in Procession over the Bridge, preceded by the Trumpets and Kettle Drums and saluted with 21 guns. On the Centre Arch was played *God Save the King* and sung by all the Company; on their Return there was another Discharge of 21 Cannon and the night was spent with the greatest Demonstrations of Joy that Men sensible of so Publick a Benefit were capable of expressing.[21]

The actual ceremony lasted for two hours but for the rest of the day Westminster was like a fairground, the river crowded with boats, the bridge packed with people passing to and fro, and the pickpockets making a day of it with a roaring trade in purses and watches.[22]

In fact the opening was generally considered to have been a success. At a meeting of the Works Committee on the Tuesday, with the Bishop of Rochester in the chair attended by Samuel Kent, Lord Luxborough and Lord Gage, Ayloffe formally announced that he had 'attended the Opening of the Bridge which was compleated by two of the Clock on Sunday Morning without any Interruption or Accident'. With the help of the other officers and Mr Neeld, churchwarden of St Margaret's, and the parish officers of Lambeth, he had stationed the watchmen and night beadles and given them written directions for their guidance:

> The Watchmen are to be at their Stands every evening as soon as the Abby Clock has struck five and are not to quit their Watch till seven the next morning.
>
> If any of the Lamps on the Bridge or Abutments thereof are broke the Watchmen are to give immediate notice thereof to Mr Walker, Tinman of Charing Cross.
>
> The Watchmen are to keep themselves sober and avoid giving offence to Passengers.
>
> They are to prevent all Nuisances and Obstructions of any kind whatsoever upon the Bridge which may annoy or obstruct the Passengers thereon and not to suffer any persons to keep Stalls or Wheelbarrows or to sell Spiritious Liquors on the Bridge and particularly not in any of the Coves or Recesses over the points of the Piers and are to take care that no loose Idle or disorderly Persons Loiter upon the Bridge or about the Recesses or Abutments thereof.
>
> They are to apprehend all persons who shall wilfully damage or destroy the Bridge or any part thereof or who shall wilfully break any of the Lamps or commit any Crime or Disorders on the Bridge and Immediately carry the said persons before some Justice of the Peace within the County or City or Liberty wherein the offence shall be committed in case the offenders shall be apprehended before Ten of the Clock at Night, otherwise they are to secure them till the next morning and carry them before the proper Magistrate.
>
> The Beadles are to see the Watch set at the hours above mentioned this Night and are to visit them frequently and see that they do their Duty and keep sober and that they report all Accidents and Misdemeanours of the Watchmen and all other necessary matters relative to the Watching of the Bridge to some one Officer belonging to the Commissioners thereof.

The watchmen were chosen by the churchwardens of their parishes and appointed by Ayloffe. A few days before the opening, the churchwarden of St Margaret's had presented Anthony Gardiner to be beadle, and Robert Thorn, Thomas Woodyer, Henry Sims, John Lockyer and Robert Williamson to be watchmen. The churchwardens of St Mary's Lambeth had presented William Page, Thomas East, Francis Smith, Charles Heather, John Oldfield and Thomas Cross. The commissioners resolved to allow ten shillings and sixpence a week to the watchmen and twelve shillings to the beadles. They were to assemble at the St Margaret's watchhouse at midnight on the Saturday, Ayloffe and the other officers allotting them their stations. Seddon was ordered to pay the churchwardens (James Neeld of St Margaret's and Peter Buscarlet of St Mary's) weekly sums of

£3 15s od and any reasonable demands for watch-coats and staves. Labelye was told to build a chimney in the carthouse near the Surrey abutment and to pave the floor with brick and rubble so that it could be used as a watch-house. [23]

A day or two after the opening, the watchmen complained to Ayloffe that they found it quite impossible to remain on duty from five in the evening till seven in the morning and the Board told Ayloffe to contrive some system of watch-keeping so that six men should be on duty at one time throughout the night. In order to make it easier Obadiah Wylde handed over to the beadles the keys of the six recesses for use as watch-boxes. They assembled every evening at St Margaret's watch-house until the Lambeth watch-house was ready, and meanwhile the churchwardens of St Mary's were authorised to hire a room near the Surrey abutment and Labelye was asked to produce drawings for a watch-house with a strong-hold beneath as soon as possible.

Damage was being caused to the superstructure even during the day-time, and two beadles, Robert Smith and John Eldridge, were appointed to keep watch during the day and to prevent stallkeepers and barrowboys from setting up in the recesses. In order to invest them with more authority Ayloffe arranged for them to be sworn in as peace officers of St Margaret's and St Mary's respectively. [24] Later on it was decided that one beadle and two labourers were enough in the daytime and after 9 March Eldridge and two men, Isaac Garner and Thomas Milton, were ordered to remain on day duty, and Robert Smith and two men, Edward Silke and Robert Jeykell, were discharged. At night during the summer the watch consisted of a beadle and three watchmen from each parish, paid at ten and sixpence and seven shillings, the wages being handed to the church-wardens as usual. [25]

The watchmen on Westminster Bridge were probably neither better nor worse than the average London watchmen, who were usually old, feeble and quite incapable of dealing with the bands of marauders roaming the streets by day and night (Plate 47). Thomas Touchit, the editor of the *Westminster Journal,* published a long article blaming much of London's crime on the incompetence of the watchmen, most of whom 'are very negligent, whence it happens that many Robberies, Burglaries and other offences, which their care might have prevented, are committed.' [26] Fielding, dilating upon the imperfections of London life and upon the administration of justice in particular, sympathises with the poor old men:

The watchmen in our metropolis who, being to guard our streets by night from thieves and robbers, an office which at least requires strength of body, are chosen out of those poor old decrepit people who are, from their want of bodily strength, rendered incapable of getting a livelihood by work. These men, armed only with a pole, which some of them are scarce able to lift, are to secure the persons and houses of His Majesty's subjects from the attacks of gangs of young, bold, stout, desperate and well-armed villains. . . . If the poor old fellows should run away from such enemies, no one I think can wonder, unless it be that they were able to make their escape.[27]

With the watchmen settled and the bridge open to traffic, all that remained was to wind up its affairs, dispose of surplus material and pay off the staff. The board ordered 'all Barges, Bargemen, Lighters and Lighter-men, Boats, Skiffs, and all waterborne People employd . . . to be discharg'd on Saturday sevenight next' and Labelye was asked to give them notice.[28] A list of the bridge officers with their various salaries had been laid before the commissioners earlier in the year and notice given that salaries would cease on 22 December.[29] Ayloffe, Lediard, Bowack and Labelye were discharged then, 'in order to lessen the expense of the publick', an instrument being signed by Horace Walpole (senior) and ten other commissioners present at the meeting. A copy of the Minutes was ordered to be laid before the House of Commons as usual and Members told that 'the several officers who have been employ'd in the Service of this Commission and are this Day discharg'd from their respective offices and employments be recommended by this Commission to the consideration of the Honble House of Commons'. Seddon was authorised to look after the commissioners' affairs and allowed £300 for himself and his clerks. He was also allowed to live in the Bridge Office house but 'no coals or candles to be allowed except such as are necessary for the Com$^{rs}$ at their Meetings'.[30] Mary Cundill the housekeeper and Thomas Lewis the clerk also stayed on. Ayloffe and Bowack had already been ordered to give Seddon 'free access to all Books, Accounts, Papers and Writings belonging to this Commission as often as he shall think fit'. Jelfe and Tufnell were ordered to build a house in Abingdon Buildings for a new Bridge Office where the various books and papers would be housed.[31]

These books were of great importance to the commissioners, recording the history of Westminster Bridge from their first meetings in the Jerusalem Chamber in 1736 to the day they were dissolved in 1853.[32] Copies still exist: thirty-five bound folio volumes in the Public Record Office, written in clerkly eighteenth-century handwriting, beautifully legible and a pleasure to read, and one hundred volumes in the House of Lords Record

Office. They were in the charge of Sir Joseph Ayloffe and were kept by Bowack, who was responsible for entering the journals, registering accounts, conveyances, and juries' verdicts, making the indexes, and seeing that the books were copied and eventually deposited in the proper public offices. This was an arduous task and in January 1741 Bowack petitioned the board for relief. He was awarded an extra £60 a year and an assistant clerk, John Lewis, at a salary of £40 a year.[33] The next year he laid before the board a ledger of extracts which gives a clear picture of the scope of his work:

> Mr Bowack, Clerk to the Committee of Accounts, laid before the Board a Specimen of a Book intituled Extracts of the Charges and Expences of building Westmr Bridge, and of all Moneys paid to Workmen, Officers and others on every part and branch of that Undertaking, and all preceeding and concurrent Contingencies and Incidents relating thereto, viz.
>
> No. 1 Contains an Account of all Expences of Solicitation, Acts of Parliament, Advertisments and Incidents preceeding the Building.
>
> No. 2 An Account of the Structure only, arising from Contracts, Bills, Allowances, and Gratuities in Sounding and Boring the River, Driving and Drawing Piles, Levelling the Foundations of the Piers, Building the Launch, Engine, Caissons, Bottoms for the Piers, and the Piers themselves, the Westmr and Surrey Abutments, the Centers to turn the Arches, the Stone Arches themselves, and Superstructure, and Incidents of various kinds.
>
> No. 3 An Account of all Contingencies, as Salaries, Office Rent, and Expences, Acts of Parliament, Advertisements, Stationery Ware, and Bank and Lottery Expences &c. concurrent therewith.
>
> No. 4 An Account of Purchases, Conveyancing, Law Charges, Fees at Office, pulling down Houses, Clearing and Rebuilding Almshouses, Wharfs, Paving, &c.

Bowack also produced a staff ledger which contained 'the particular Charges and Discharges of all persons employ'd by the Commissioners'. The two books were examined by the board and Bowack was ordered to keep them up to date.[34] Unfortunately both these ledgers have disappeared but the surviving thirty-five volumes contain enough information, especially in the details of land conveyance for the approaches, to provide a valuable source book of London topography.

Towards the end of each year Ayloffe was called to the Bar of the House of Commons to give an account of the progress of the new bridge. He presented the Minute Books for the year, the Contract Books and the Treasurer's Accounts, and then withdrew so that the Clerk of the House could read out the titles. The books then lay on the Table to be perused by Members and afterwards to be preserved among the sessional papers.[35] For the last few years Ayloffe handed on this duty to Bowack and Samuel Seddon. It was reported baldly in the Commons Journals. The more interesting details were given to the House in the reports, usually presented by one of the Members for Westminster, following the commissioners' petitions for further powers and more money.

The disposal of surplus stone, timber and ironwork was soon settled. A notice was published advertising the sale of rubble and stone lying near the abutments and buyers were told to hand in their proposals by 15 January.[36] Two hundred tons were bought by Henry Cheere, the sculptor, for use of the new Westminster Fishmarket, of which he was Comptroller to the Trustees; and the remainder about 330 tons, was bought by Thomas Phillips at three shillings a ton and resold to the Fishmarket.[37]

The board accepted an offer made by Joseph Lydall, a carpenter of St George's Hanover Square, to buy all the old ironwork at eleven shillings a hundred and a few weeks later Obadiah Wylde reported that most of the old iron, brass and junk had been sold to Lydall, £61 1s 6d on account being paid to the treasurer. Surplus stores and tools were sold off for £33 13s 0d and by March 1751 all that remained were 'the Centers under the 68ft and 64ft Arches with their Puncheons &c., 6 Stage Piles below the sunken Pier, the Center under the Counter-Arch, the Pile Driving Engine compleat, the Saw Engine with a large Float, a House Boat, a large Wherry, and 6 Pieces of old Floats'. Joseph Lydall bought these, too, for £103, and agreed to remove them before 24 June. In April Labelye and Simpson reported verbally to the board that they had examined the centres under the two arches and the centre under the counter-arch and reckoned they could be sold for about £100, except for sawing the piles still in the river and weighing the puncheons. The board agreed and ordered them to arrange the sale and pay the money to the treasurer.[38] A few days later an agreement was signed with Joseph Lydall for the purchase of the three centres and the accompanying piles and booms for £103, the underwater piles to be cut and the puncheons to be weighed at the cost of the commissioners, the work to be finished before 24 June.[39]

Finally Seddon was directed to settle up with John Smith of Smith & Ransom's yard at Stangate for any balance due for the use of cranes, the dock and wharf, and for any damages. They agreed on £323, thus ending a very amicable business arrangement lasting for over twelve years.[40]

One other obligation was due from the commissioners. During the course of construction the bridge had taken its inevitable toll of life and limb from the workers, most of the accidents being duly reported and compensation paid to the dependants. The casualties were received and patched up by the Westminster Infirmary in James Street, the nearest hospital to the site, and shortly before the opening ceremony the infirmary trustees decided to present a petition to the commissioners pointing out that during the building of Westminster Bridge various accidents had happened, sometimes as many as three a day and often with fatal results. The Westminster Infirmary, they said, was the nearest hospital able to cope with the casualties, which had all been treated promptly, and the treasurer, the Reverend Pengry Hayward, enclosed a sample list of patients admitted during the years 1749 and 1750.[41] The infirmary relied for its maintenance on charitable contributions and the trustees' hope that the Bridge Commissioners would help towards the good work was met with an award of £100 'in consideration of the many cures performed upon Workmen Labourers and others who have been Lamed and Wounded in carrying on the Building of the Bridge'.[42]

The infirmary's estimate of as many as three accidents a day was probably an exaggeration designed to add colour to the petition. But certainly accidents did happen and were recorded in the Minutes. The work was dangerous and accidents are an occupational hazard of bridge building. Batty Langley was quick to point this out at the time of the sinking pier, smugly hoping that 'the present unhappy Misfortune . . . may be the last and be surmounted without Loss of Lives and Limbs which is very much to be feared, the taking down of the dislocated Arches being a very dangerous Work'.[43]

One of the last accidents to happen on the bridge before its completion was when Horace Walpole's chairman, Robert Plumridge, fell off the balustrade and was drowned. What he was doing on the balustrade we are not told, but his widow Martha Plumridge appealed to the commissioners for help. The board signed a warrant there and then, directing Seddon to pay her ten guineas.

## NOTES TO CHAPTER 14

1. Laroche, Stanhope, Horace Walpole (junior), Hales, Salisbury, Glanville, the Bishop of Rochester, Kent, Ellys, General Oglethorpe, Admiral Warren, Mr Belcher and Pembroke himself.
2. Bridge Minutes 9 January 1749/50.

3. Horace Walpole to Horace Mann 10 January 1749/50.
4. Bridge Minutes 6 March 1749/50.
5. *London Magazine* (1750) p 39.
6. CJ 17 February and 6 March 1749/50, vol 25 p 1031.
7. Bridge Minutes 16 January 1749/50.
8. Ibid 13 March 1749/50.
9. Ibid 13 and 22 February 1749/50.
10. *Gentleman's Magazine* 8 February and 8 March 1750 pp 89 and 137; *London Magazine* (1750) pp 91, 138–9; C. Davison *History of British Earthquakes* (1924) pp 333-6.
11. Labelye *Description* p 22.
12. *British Museum Catalogue of Political and Personal Satires* 3076.
13. Horace Walpole to Horace Mann 2 April 1750.
14. Works Minutes 29 March 1749/50; Bridge Minutes 10 April 1750.
15. Labelye *Description* pp 84–5.
16. Bridge Minutes 1 June 1750.
17. Works Minutes 17 July and 28 August 1750.
18. Bridge Minutes 6 November 1750; *General Advertiser* 7–9 November 1750.
19. *London Magazine* (1750) p 523.
20. Bridge Minutes 13 November 1750; *London Magazine* (1750) p 523.
21. *Whitehall Evening Post* 17–20 November 1750, repeated more shortly in other papers; see also *Gephyralogia* p 115.
22. *Penny London Post* 19–21 November 1750; *Gentleman's Magazine* (1750) pp 523 and 586.
23. Works Minutes 20 November 1750.
24. Ibid and Bridge Minutes 13 November 1750.
25. Bridge Minutes 5 March and 16 April 1751.
26. *Westminster Journal* 2 April 1748; *London Magazine* (1748) p 177.
27. Henry Fielding *Amelia* (1751) Book 1 Ch 2.
28. Bridge Minutes 23 October 1750.
29. Ibid 27 March 1749/50.
30. Ibid 14 December 1750.
31. Ibid 17 April 1750.
32. See Bibliography page 296.
33. Bridge Minutes 28 January 1740/1.
34. Ibid 9 and 30 June 1742.
35. See for example CJ 9 December 1746, vol 25 p 209.
36. *General Advertiser* 17, 18, 19 December 1750.
37. Bridge Minutes 7 and 14 May 1751.
38. Ibid 12 February, 5 March and 16 April 1751.
39. Ibid 23 April 1751.
40. Ibid 12 March 1750/1.
41. Weekly Board of the Publick Infirmary in James Street, 7 November 1750.
42. Bridge Minutes 11 December 1750; *General Advertiser* 21 December 1750.
43. Batty Langley *A Survey of Westminster Bridge* (1748) viii.

# 15
# Westminster Bridge, 1751

Early in 1751 the governor and company of Chelsea Waterworks, through their clerk Timothy Brent, applied to the commissioners for compensation for decrease in water rent due to the demolition of numerous houses in Westminster. The board decided that this would not be proper but they did agree to compensate the company for the loss of their water pipes and ordered Seddon to inspect the company's books. At the next meeting Seddon reported that he calculated the waterworks had lost 912 yards of piping buried in the ruins of 152 houses, which at the current price of three shillings and fourpence a yard amounted to £152. Seddon was ordered to pay the company £150.[1]

At the next meeting the board interviewed Daniel Gell, the agent for the Dean and Chapter of Westminster Abbey, and then passed on to the main topic of the day, which was the question of an honorarium for Labelye, who for twelve years had served the commission at a salary of £100 a year, fixed when he had been appointed engineer in 1738. A week or so after the opening ceremony, in a memorial addressed to the commissioners, Labelye had congratulated them on the completion of their great work, declaring that 'to their Publick Spirit, their constant Care & singular Disinterestedness, the Publick chiefly owes not only that there is a *Bridge at Westm^r*, but that it is a Stone Bridge, the largest that ever was built over a Tide River'. He praised them for the methods they had chosen to use, taking care to point out that the fourteen piers, mostly set up in deep water, had been built to his own design after many other methods had been examined and rejected. He touched on his own impartiality in recommending to the board the various engines and inventions put forward by Vauloué, King, Etheridge and others, and expressed his appreciation that they had all been rewarded 'not only with Reputation but with suitable and handsome Gratifications'. But what gave him most satisfaction was 'that notwithstanding all the false, malicious and scandalous Reports rais'd and artfully propagated to depreciate the Works of the Bridge, or to lessen the Characters of the Persons in anywise con-

31. *Lord Mayor's Day 1746* by Canaletto. (*Paul Mellon Collection. Photograph by courtesy of the Paul Mellon Foundation*)

32. Detail of Plate 31. Thames and Isis.

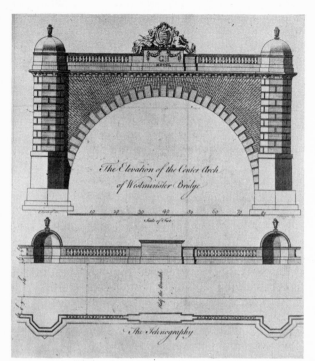

33. Elevation of the middle arch with the river gods, Thames and Isis. (*Drawing by Gruyn engraved by Rooker for* The Gentleman's Magazine *1754. Photograph by courtesy of the British Museum*)

34. *View through an Arch 1746–7 by Canaletto.* (*Royal Collection reproduced by gracious permission of Her Majesty the Queen. Photograph by A. C. Cooper*)

cerned therein, those Persons have now the satisfaction to find that Westm<sup>r</sup> Bridge is seldom seen or mention'd, even by its greatest Enemies, without some commendation'. The memorial ended with the humble hope that the commissioners had not been disappointed in their expectations of him. [2]

The board fully appreciated the implications of Labelye's memorial, resolving to consider it at a future meeting, but although there were several meetings in January and February it was not discussed until 26 February with Horace Walpole (senior) in the chair and twelve other commissioners present. [3] After they had dealt with the affairs of Daniel Gell and the Chelsea Waterworks, they turned to Labelye again. The engineer had become impatient and had followed up his memorial with another, even longer and more detailed and putting the point at issue with more emphasis. He claimed with every justification that considering the responsibilities he had shouldered over twelve years of his life, the annual allowance he had received from the commissioners (£100 plus expenses) amounted to no more than a pittance. An examination of the accounts shows that he had been paid £25 quarterly with ten shillings a day sub-sistence amounting to about another £45 a year. Various small disburse-ments were also entered to his account. In 1740, for instance, he had been paid £8 18s od for several drawings, £26 19s od for models of a stone arch and a centre, and £15 2s 6d for travelling expenses to the Dorset quarries. [4] Ayloffe's and Blackerby's salaries were £200 a year each, Graham's £300 and the two Lediards' £300 with occasional £100 bonuses, and Labelye had every right to feel that he had not been paid his due. His second memorial, therefore, laid before the commissioners in February 1751, is really a model of restraint, especially when allowances are made for English not being his native language.

> To the Right Honble &c the Commissioners of Westminster Bridge
> I do humbly beg leave (on this occasion) to lay before the Board the following short account of the time I have been in their Service. I was appointed their Engineer in May 1738 to direct the laying the Foundations and building the Piers, by a method of my own, which appeared to the Board the cheapest and most practicable out of all the Schemes offered to them; by which I have saved this Commission the vast sums which the attempting to do this by Coffer Dams or any other known methods would have cost, and perhaps the Disgrace of not succeeding at last—
> I was allowed then 10 Shillings a Day and £100 a Year for Subsistence, and was promised from the Chair, by Order and direction of the Board, that in Case my method of building the Piers should succeed, the Commis-sioners would amply reward me and provide for me in a handsome Manner

for Life. I have in consequence directed the Building of the 14 Piers of the Bridge which were all finished in less than 4 years, with all the success imaginable and at an expense very moderate. In April 1740, when the Wooden Superstructure was laid aside and a Stone Bridge resolved upon, I gave the Design according to which it is now finished. And by a new Commission the Board appointed me to direct the Building not only of the Piers but of the Abutments, Arches, Superstructure, and all the Works relating thereto. Though my Province, Duty, Care and Attendance were then much enlarged, no addition was then made to my Allowance and I asked for none, relying on the promises made me at first and the honour of the Commissioners. I have now the pleasure to see the Bridge the best built and the finest in the World compleatly finished under the orders of the Board and my direction, and I hope to the general satisfaction of the Publick, without turning off any Part of the River, without hindering or stopping the Navigation of the Thames one single moment, and without occasioning any Sensible Fall under the Arches.

During the whole time of Building I have constantly attended the Works, both by Day and Night, doing myself what is usually done by a clerk, which is commonly allowed to an Engineer, but what I never so much as asked for —I have saved this Commission considerable sums on almost every article by a proper choice and Disposition of the materials, and in Particular upwards of Five Thousand Pounds in the single article of Centers.

Besides the Compleatly building the Bridge and successfully repairing the Damages it sustained by the Settling of one of its Piers, I contrived and directed the making of the New Road in Surry, I took the Levels and laid out the New Streets for Mr Lediard Senior, and assisted him in building the Sewers under them and raising them as they are now by the Directions of the Board. I made all the Surveys and original Maps for the New Streets and Surry Road, and also all the Draughts and Estimates for the use of the Commissioners and their Artificers, and in particular all the many duplicate copies of the Draughts which were annexed to the Contracts—for all which things (besides my Planning, directing and attending the Works of the Bridge) I never was allowed anything, nor did I ever charge or demand anything, not even for Stationery wares, the Imperial Paper only which I have used amounting to a considerable sum.

I never had any perquisite or Fee whatsoever from any of the Masters or Artificers or any other Person whatsoever, nor ever expected or demanded any. On the contrary I avoided designedly (for fear of any Imputations of this sort) to meddle in the Contracts. I also contrived that not one single Shilling of the Commissioners' or the Publick Money should ever pass through my hands, except Allowances and a few trifling Disbursements for a Waterman's Wages and a Labourer on the New Road, by order and at prices fixed by the Board. In these several Works I have endeavoured to prove myself not an unprofitable servant to this Honble Commission in whose service I have spent above 12 Years and the Prime of my Life, and greatly impaired my Health having contracted Asthma (daily growing upon me) by my constant attendance on the Works, especially on the Water in

Winterly and Rainy Weather. Savings out of my Allowance I own I have made but little or none, and I have refused more than once considerable Offers from abroad, relying entirely on the Promise made to me at first, and the Honour of the Commissioners; all which is humbly submitted to the Consideration of this Hon^ble Board

By their most humble Servant

Westminster                                    Charles Labelye
February 26th 1750–1

The commissioners considered Labelye's two memorials together and unanimously agreed that he deserved a generous reward over and above his usual allowance. 'For his great Fidelity and extraordinary Labour, Attendance, Skill and Diligence in performing all the Services . . . under this Commission', they awarded him a gratuity of £2,000. They also resolved that application should be made to Parliament for authority to continue his employment with the commission at a yearly salary of £150.[5]

This was handsome enough and Labelye acknowledged their generosity a few months later when he published his last pamphlet on the bridge, *A Description of Westminster Bridge* (1751), a work ordered by the board at the end of the same meeting. Some years earlier he had actually planned to write a book on Westminster Bridge in two volumes, and in July 1744 had written a preface and table of contents which were appended to *A Description*. Ill health and possibly other troubles intervened and the work was never done. It was to have been called 'A Description of Westminster Bridge to which are added Historical Accounts relating to the Building and Expence thereof with technical Descriptions of all the Operations, Machines, Engines, &c. made use of in the course of the Works. Also Analytical Investigations Calculations and Geometric Constructions referr'd to in the former Parts, with Practical Rules and Observations in Mechanics Hydraulics and the Art of Building in Water'. He planned to illustrate the book profusely with copper plates and the title page was to have been adorned with the lines from Pope's *Epistle to Lord Burlington,*

> Bid Harbours open, public Ways extend;
> Bid Temples, worthier of the God, ascend;
> Bid the broad Arch the dang'rous Flood contain,
> The Mole projected break the roaring Main:
> Back to his bounds their subject Sea command,
> And roll obedient Rivers thro' the Land:
> These Honours Peace to happy Britain brings,
> These are Imperial Works, and worthy Kings.

He possibly intended his book to rival the textbook which no doubt he used during the course of building the bridge, *Le Science des Ingénieurs,* published in Paris in 1729 by Bernard Forest de Belidor, a technical officer in the Corps des Ingénieurs des Ponts et Chaussées and a teacher of mathematics and physics at the Artillery College of La Fère. This was one of the earliest concise manuals in civil engineering and contained the scientific data which a bridge builder in the 1730s and 1740s would certainly have needed. Belidor's other work, *Architecture Hydraulique,* published in four volumes between 1737 and 1753, is of the greatest interest for the insight it gives into the details of mechanical aids available at the time: suction and pressure pumps, dredgers, a primitive steam engine called Papin's *machine à feu,* and other devices immediately preceding the Industrial Revolution. Labelye's two volumes would certainly have contained a great deal of material not mentioned in the commissioners' journals. As one historian said at the time, 'it is to be regretted that the ingenious and honest gentleman who was appointed over this work did not, at the finishing of it, publish all the particulars of the mechanism of this grand structure, in a manner he was capable of doing, to the satisfaction of the curious'.[6]

Labelye was naturalised by Act of Parliament in 1746. There is a mystery about where he lived in London. In 1739 he lodged in Derby Court. According to Telford, who lived at 24 Abingdon Street, Labelye had lodged in the same house, where there was a Canaletto painting of the bridge over the dining-room fireplace.[7] The Westminster rate books provide no clue; up to 1752 Abingdon Street was rated as Lindsay Row in the Grand Division of St Margaret, and both Colonel Tufnell and Thomas Lediard lived there. Possibly Labelye lodged with one of them and hoped to write his book under their roof. But in April 1751 he lamented the death of Pembroke and other commissioners who had been his friends, and though 'the Commissioners according to their Promises and without any Solicitations from me, have been pleased to afford me both *Leisure and Means* sufficient to enable me to go on with it', his health was failing and the book was never done.

From this point Labelye disappears almost without trace. When the time came, in September 1751, to saw off the last piles and weigh the puncheons, the surveyor 'went several times to Mr Labelye's lodgings but was always told he was out of Town and that it was not known when he would return'.[8] On 5 April 1752 he wrote a letter to Seddon certifying Jelfe and Tufnell's account for making a drain and for chipping ('tooming') the pier and springers after the bridge had been opened.[9] On 9 April he called at Seddon's house and asked him to give the commissioners a message:

> He being greatly afflicted with an Asthma and unfit for Business, was going to the South Part of France for the recovery of his Health; that he proposed residing at a small Town there called Béziers for about a Twelve Month, and that he would not during that time engage himself in any Foreign Service, but wait there till after the next Session of Parliament to know whether he should be wanted to be employed here, and that in case he should not be wanted, he hoped he might afterwards, without Reproach, be considered as his own Master and be at Liberty to Dispose of himself as should be most for his conveniency or to that effect; and that in case he should be wanted he left the following Directions with Mr Seddon how he might be sent to, viz^t
>
> Letters to Mr Charles Labelye to be sent undirected to Mr John Robertson, Master of the Mathematicks in Christs Hospital.
>
> And in case of Mr Robertson's death, to Messrs Henry and Joseph Guinand, merchants in Little St Helens, Bishops Gate Street, London.

The cure at Béziers seems to have been successful. He returned to Paris and met Perronet to whom he bequeathed some papers and a model of Westminster Bridge, though this has since disappeared. Even the date of his death is uncertain. He probably died in Paris in 1782.[10]

The happy outcome of Labelye's application to the commissioners resulted in a stream of letters from the other officers, some of whom had evidently not availed themselves of the opportunities for personal profit which their position on the bridge would have made quite normal at that time. Lediard in particular seems to have managed his affairs without much financial success. His father had died insolvent in 1743, having invested heavily in land on the bridge approaches, and Lediard junior had been unable to free himself, in January 1747 being in debt to the commissioners for over £400. He petitioned for time to pay, claiming that he had served the bridge 'upwards of 3½ years and humbly hopes with Fidelity & Diligence, & that he cannot be charged with having defrauded the Trust of a single Farthing'. Most of the debt had been caused by his being obliged to finish the house begun by his father on bridge land 'for the encouragement of builders . . . and lately assigned over to the workmen at the loss of upwards of £500'. Lediard ended his petition with the plea that the commissioners would consider his case '& not take any steps for removing him the rumour of which only would inevitably end in the ruin of himself & a large Family who have no other dependance'. The board allowed him a fortnight to pay and with the help of an arrangement for the payment of rent the debt was settled.[11]

But then, with the completion of the bridge and the termination of the officers' employment, Lediard was compelled to petition the board again on the grounds that

> Your Petitioner's Father, in order to encourage builders to take ground on the Abingdon Estate, did take a piece of Ground himself and began to build a House which your Petitioner was oblig'd to finish and by which your Petitioner lost a considerable sum of money.
>
> That your Petitioner's Father having a very numerous Family and being many years out of Employ died insolvent and yr Petitioner out of regard to his memory paid several Debts he owed, particularly a large Debt to your Honours, by means of which, the finishing of the Building and the maintaining his Father's Family, yr Petitioner is reduced to such unfortunate circumstances as not to be able to support himself, & Mother & 3 Sisters, now he is discharged by yr Honours.

Lediard's petition was effective enough for the board to resolve to apply to Parliament for £12,000 to carry on their affairs and from which a year's salary each should be paid to Ayloffe, Lediard and Bowack.[12]

Bowack presented a memorial in March 1751 showing that he too had served the bridge faithfully since his appointment as assistant clerk in 1736 (and before that had arranged the Horn Tavern meetings), that the work had increased so much that by 1740 he had eight large folio volumes to keep, that he had been given a clerk to help (John Lewis) and had received several gratuities to compensate for 'his extraordinary Labour Pains & Expense'. But the clerk had died in June 1748 and from then on he had had no help

> being forced to labour almost day and night himself in a very uncommon manner & to get what assistance he could at a considerable expence, to copy the Proceedings for Parliament, & dispatch the Business of the Office in due time, by which Fatigue he greatly impaired his Health & Sight, & so broke his continuation that in April last he was forced to apply to this Honble Board for leave of absence for some short time (which with great humanity was granted) . . .

For the last two and half years, continued Bowack, he had had daily recourse to over twenty large folios and hoped that the commissioners would duly consider his constant attendance. The board did consider, and awarded Bowack £110 for his extraordinary services and expenses.[13]

The treasurer, Samuel Seddon, appears not to have petitioned. Nevertheless the commissioners took into consideration the fact that no allowance had been paid him from 23 December 1746 for collecting the rents.

His accounts showed that he had collected £1,491 4s 7½d and they resolved to pay him at the rate of one shilling in the pound, making an allowance of £74 11s 0d. [14] Seddon seems to have benefited more than anyone. He had been appointed treasurer after Nathaniel Blackerby's death in 1742, earning £120 a year until the pay-off in December 1750, when he had been put in complete charge of the bridge affairs at a salary of £300, and with the use of the Bridge Office as his home.

With the bridge open, in constant use and unhampered by tolls or pontage, the commissioners' work became noticeably easier, consisting mainly of routine care and maintenance. In March 1751, petitioning in the usual way for more powers and money, they announced that although they had 'with the utmost care, caused the Bridge to be built and made passable', they were still unable to 'make an effectual and constant provision for maintaining, supporting, cleaning, watching, lighting, paving and gravelling the said Bridge and for preserving it from Annoyances and other Damages.' Laroche then reported from the committee that the engineer had informed them the bridge had been opened on 18 November and the works had been carried on with the greatest frugality—one of Labelye's favourite phrases. Seddon had also reported that most of the approaches to the bridge had been finished three years ago. The House recognised that more powers were needed, and leave was given to bring in another Bill 'to enable them more effectually to discharge the Trusts reposed in them, and for making an effectual and constant Provision for watching and lighting the said Bridge and for preserving it from Annoyances and other Damages'. [15]

During the daytime all that was needed to maintain the bridge was the watch to prevent annoyance to passengers and damage, and a gang of men to keep the surface in repair. There was an efficient drainage system which coped with rain-water, but in the summer traffic over the sand and gravel surface caused clouds of dust to rise from one end to the other. The road-maker, John Simpson, was ordered to keep the bridge watered on dry days, and contracted with a carman, Thomas Cosby, to provide horses, cart and men to pump and water the bridge for seven shillings a day. After working for several days during the next fortnight Cosby attended a board meeting to say that he could not continue watering at that price, and 'after some Discourse' they agreed on seven shillings and sixpence 'for Watering the Bridge on Dry Days'. [16]

At night the bridge was lit by thirty-two oil lamps maintained by a tin-man of Charing Cross, Thomas Walker, who had agreed to make square lamps, according to pattern, of the strongest double tin at three guineas each, and globular lamps at two guineas. He had also agreed daily 'to light each lamp containing three Spouts from Sun Setting to Sun Rising' for five pounds a year, and 'to keep the whole in Repair and painting the Iron Work & replacing any number under 5 that shall be broke at any one time'.[17] The lamps had been provided by a blacksmith, Benjamin Holmes, who in 1747 had been ordered to make 'a pattern lanthorn for the use of the Bridge agreable to the Drawing'.[18]

Buck's view of Westminster Bridge, published in September 1749, shows the bridge complete except for the lamps. They were not actually placed in position on the balustrade over the recesses until just before the opening ceremony, Holmes excusing himself for not having them ready earlier but promising them for the occasion.[19] Once they were up they immediately became a target for vandals, 'several being of late wilfully and maliciously broke', and the usual penalty was threatened—death as a felon without benefit of clergy—and a five-guinea reward offered to anyone securing a conviction.[20]

Normally the bridge was brightly lit at night and from November onwards throughout the winter three spouts and burners were used, the tinman being paid £72 2s od for lighting and maintaining the thirty-two lamps on the bridge and its abutments from 18 November 1750 to 30 April 1751.[21] Walker was taken to task by the commissioners on one occasion because the lamps had not been lit between seven and eight o'clock one evening. He gave as his reason 'the Weather being then stormy, which occasioned his servants being much longer in getting the Lamps lighted at that time than they usually are in calm Weather'. The board accepted this but in order to reduce the expense of watching and lighting during the summer they instructed Walker, for a week's trial, to light one burner only on each lamp from eight o'clock.[22] Walker's men can be seen lighting the lamps in Scott's 'Arch of Westminster Bridge' in the National Gallery of Ireland.[23] Thomas Walker died in 1753 and the lamplighting on the bridge was continued by his widow, Margaretta Innocentia, and her brother Jonathan Durden.[24] Since the Watching and Lighting Acts of the 1730s and 1740s the street lighting in London had improved considerably, until the Prince of Monaco, visiting George III in 1768, thought the town had been specially illuminated in honour of his arrival. And it was said that there were more lamps in Oxford Street alone than in the whole of Paris.[25] Gaslights were fitted on Westminster Bridge in 1814 and the old methods were looked back on in scorn.

Forty years ago the lighting of the streets was affected by what were called 'parish lamps' [says a Victorian writer]. The lamp consisted of a small tin vessel, half filled with the worst train oil that the parochial authorities, for the most part chosen of the select vestries, could purchase at the lowest price to themselves and the highest charge to the rate payers. In this fluid fish-blubber was a piece of cotton twist which formed the wick. A set of greasy fellows redolent of Greenland Dock were employed to trim and light these lamps, which they accomplished by the apparatus of a formidable pair of scissors, a flaming flambeau of pitched rope and a rickety ladder, to the annoyance and danger of all passers-by. The oil vessel and wick were enclosed in a case of semi-opaque glass . . . which obscured even the little light it encircled. [26]

Walker and his stinking lamps are even immortalised in the poems of W. M. Praed:

> And hammer out the tedious toil
> By dint of Walker and lamp-oil. [27]

Labelye estimated the total cost of the bridge to have been in the region of £218,800, which although undoubtedly a far cry from the original estimate of £90,000 for a timber bridge put to the commissioners in 1739, he justified in a description of 'the remarkable Frugality which the Commissioners have shown all along in laying out such sums of publick Money'. [28] He instanced as examples of this economy the spandrels and secondary arches being made of Purbeck stone, the timberwork being usually of soft wood (oak being used only where absolutely necessary), and the centres being designed to fit the corresponding arches on both sides of the bridge, much to the annoyance of the timber merchants, who had reckoned on separate centres under all the arches. Labelye declared that the Earl of Pembroke had often told him that even at £300,000 Westminster Bridge would have been cheap. And when the Society of Gentlemen were holding their meetings in 1734 and 1735 the supporters of a timber bridge had considered that a stone one would have taken twenty years to build and would have cost from £400,000 to £500,000; whereupon a stone bridge supporter had replied that even if it did take twenty years and cost £500,000, 'it was in every way preferable and much more suitable to the Place than any wooden Superstructure whatsoever'. [29] Labelye claimed that the twelve years and £218,800 actually spent on the bridge were an achievement of which the commissioners could be proud. He compared it to St Paul's Cathedral, which had taken thirty-five years to build, had cost £736,800 according to Maitland's *History of London,* and contained about half the quantity of stone used in Westminster Bridge. Moreover, the cathedral's foundations had not been laid in water, and its masons had not had to work against two high tides every day. A later calculation, based on the lottery takings and state subsidies, estimated the cost to have been £389,500.

## NOTES TO CHAPTER 15

1. Bridge Minutes 26 February and 5 March 1750/1.
2. Ibid 14 December 1750.
3. Horace Walpole, Viscount Fauconberg, Lord Archer, the Bishop of Rochester, Sir Francis Poole, Sir John Heathcote, Sir Thomas Hales, Sir John Elwill, Sir Peter Thompson, John Laroche, Samuel Kent, Henry Fox, William Hay, William Clayton, Colonel William Herbert, Robert Fairfax, Thomas Salusbury, General Oglethorpe, William Turner and Horace Walpole (junior).
4. Bridge Accounts pp 22, 58, 85.
5. Bridge Minutes 26 February 1750/1.
6. Stephen Riou *Short Principles for the Architecture of Stone Bridges* (1760).
7. Joseph Mitchell *Reminiscences of my Life in the Highlands* (1883) pp 87–8 describing life with Telford at 24 Abingdon Street.
8. Bridge Minutes 19 November 1751.
9. Ibid 7 April 1752.
10. *Dictionary of National Biography*.
11. Bridge Minutes 2 January, 3 and 10 February 1746/7.
12. Ibid 5 March 1750/1.
13. Ibid.
14. Ibid 19 March 1750/1.
15. CJ 6 March and 2 April 1751, vol 26 pp 94 and 163–4. Laroche, Kent, Horace Walpole, Hales, Burrell, Hay and Cooke were given leave to prepare the Bill.
16. Bridge Minutes 4 June 1751.
17. Ibid 24 April 1750: Bridge Accounts 19 December 1750.
18. Bridge Minutes 31 March and 17 November 1747; Works Minutes 1 May and 17 July 1750. Holmes's bills for blacksmith work on the bridge between 1738 and 1750, including repairs to the pile-driving engine, came to £1,990 13s 4d (Accounts pp 25 and 59).
19. Bridge Minutes 23 October 1750.
20. *General Advertiser* 17 December 1750.
21. Bridge Minutes 23 and 30 April 1751.
22. Ibid 16 April 1751.
23. This painting by Samuel Scott is a slightly later version, showing the iron lampstands, of the Scotts in the Mellon Collection and the Tate Gallery (Plate 42).

24. Bridge Minutes 10 April 1753 and 19 March 1753/4.
25. Archenholtz *A Picture of England* (1797) p 131.
26. J. Richardson *Recollections of the Last Half Century* (1856) quoted by M. D. George *London Life in the XVIII Century* (1925) p 102.
27. W. M. Praed *A Preface* (1824).
28. Labelye *Description* pp 88–97.
29. Ibid p 94.

# 16
# Paintings of the Bridge

In 1781 the first clerk to the commissioners of Westminster Bridge, Sir Joseph Ayloffe, died at his home in Providence Court, Westminster. Ayloffe had become a distinguished antiquary. He had been for many years vice-president of the Society of Antiquaries and among his many contributions to the Society's journal, *Archaeologia,* was an account of a painting at Cowdray Park representing the coronation procession of King Edward VI. It was a panorama taken from the north, probably from Highgate, and showed the procession leaving the Tower of London and passing along Eastcheap, Gracechurch Street, Cornhill and Cheapside, and by means of an ingenious contraction of the time scale, arriving at Westminster for the culmination of the ceremony, the coronation in the Abbey itself. Behind the procession can be seen the Tower, the Temple and old St Paul's, and in the background the river Thames flowing peacefully under London Bridge and curiously confined along the Southwark shore by a neat brick embankment. The artist was generally thought to have been Holbein, who had stayed at Cowdray Park with Sir Anthony Brown, but Ayloffe doubted this attribution and as an alternative suggested the name of Theodore Bernardi (or Lambert Bernard) who had painted historical pieces in Chichester Cathedral in 1519 and afterwards had settled with his family in that part of Sussex. Ayloffe mildly reproached Horace Walpole for describing the 'Procession' as by no means a work of art, but a mere curiosity. In this sort of painting, argued Ayloffe, the artist must confine himself strictly to the facts: 'his landskip is to be the real face of the country... and the buildings such and such only as then stood thereon'.[1]

This panorama was among a group of historical paintings in the lower dining room at Cowdray Park, destroyed in a disastrous fire in 1793. Luckily they had all been described in some detail by Vertue in 1751 and by Ayloffe in 1773, and drawings had been made by S. H. Grimm for the Society of Antiquaries, in whose library they still remain. The coronation procession, known sometimes as the 'Riding from the Tower', is familiar from a large engraving made by James Basire. A drawing in the Ashmolean

Museum, known as the 'Long View of London from Westminster to Greenwich',[2] and the 'Riding from the Tower' are the forerunners of a line of topographical views of London and the Thames culminating in the paintings of Samuel Scott and Canaletto. The 'Riding from the Tower' was taken from a definite viewpoint on the high ground to the north of London, but this was exceptional and in most subsequent panoramas the artist tried to imagine himself viewing the scene from an angle high in the air—in fact, bird's-eye views or prospects. Prospects were quite distinct from the strictly two-dimensional maps and surveys, though in the main both categories were produced by Dutch and Flemish artists. English artists did not readily take to this genre of painting until well into the eighteenth century and the spirit of the Thames was most eloquently evoked by men such as Claude de Jongh, Cornelius Bol and Abraham Hondius, all well accustomed to the atmospheric haziness of canal life in the Low Countries.[3] Peter Monamy, a Channel Islander, lived at Westminster 'near the River side, for the Conveniency in some measure of viewing Water & Sky'.[4]

The tradition was continued, perhaps rather less poetically, by Leonard Knyff and Jan Kip and by a family of Dutch painters who settled in London after the Fire, Jan Griffier and his two sons, Robert and Jan Griffier II. According to Horace Walpole, the elder Griffier 'bought a yatch [sic], embarked with his family and his pencils and passed his whole time on the Thames between Windsor, Greenwich, Gravesend &c'. He died in 1718 at his house in Westminster at the age of seventy-two. The first known view of Westminster Bridge, painted by his son Jan Griffier II, is really a prospect of the Thames during the Frost Fair of 1739 and shows the curve of the river at Whitehall with St Paul's, the Tower of London and the new City churches in the background (Plate 18). Interest is centred on the booths and tents in the foreground and on the foot-track across the ice from Whitehall Stairs to Lambeth Marsh just above the notorious Cuper's Gardens. A subsidiary point of interest is marked by the two middle piers of Westminster Bridge, with sight-seers threading their way across the frozen river to inspect the works and then continuing along a line of booths following the course of the bridge to the wharfs and timber yards on the Surrey shore. The artist probably painted the scene from an upper window of Montagu House, though the effect is of a bird's-eye view from a point even higher. The horizon is set high on the canvas, the great boulders of ice give an illusion of clouds reflected in tranquil water, and the innumerable figures scurrying to and fro still have a seventeenth-century Netherlands quality.

Several years earlier than Griffier's view, an English artist was at work on the Thames—Samuel Scott, known loosely as 'the English Van de Velde' or 'the English Canaletto', though neither of these labels means very much, or does justice to his painting. His early work was certainly influenced by the Van de Veldes and in a remarkable series of pictures of settlements on the East India Company's shipping route to the Far East, he showed himself a marine painter very much in the Dutch seventeenth-century tradition. In this particular series Scott combined with George Lambert, painting the shipping himself and leaving the landscape and buildings to Lambert.[5] The two artists received their commission in about 1730 and the bill was paid and the pictures were hanging in the Company's Leadenhall Street offices in 1732. By this time Vertue thought Scott to be among the best artists working in London, 'where Art flourishes more than probably it has done 50 or 60 years before—in Numbers of artists & works done, which in great manner is owing to the peacefull times & travailling thro Europe Italy or Rome &c.'[6] Vertue also noted that Sir Robert Walpole had seascapes by Scott over the doors of the Levée Room in Downing Street and landscapes by Griffier over the doors of the Green Velvet Bedchamber.[7] Sir Edward Walpole, the Prime Minister's second son, was also a keen collector of Scott's work. Curiously enough the younger son, Horace Walpole, though an admirer of Scott's painting, did not honour him with a place in the first edition of the *Anecdotes of Painting* (1762) but praised his work enthusiastically in later editions and in a letter to Horace Mann in 1771.

Meanwhile Scott was producing paintings and drawings of the Thames from Deptford to Twickenham and Richmond. An endearing charm was lent to this activity by the famous five-days' spree to Sheppey, Scott accompanied by Hogarth, Thornhill, Ebenezer Forrest and William Tothill, setting out on a Saturday morning from Covent Garden to the tune of 'Why should we quarrel for riches', and after a series of pranks and escapades arriving back at Somerset Watergate on Wednesday afternoon. The journey was written up as a parody of an antiquarian itinerary and exquisitely illustrated by Hogarth and Scott.[8]

The building of Westminster Bridge immediately captured Scott's imagination. As the piers were sunk and the arches rose across the river he recorded the work in minute detail, though strangely enough the first major painting of the subject was not made until the spring or summer of 1742, when the two middle arches had been turned, the timber centre for another (east 72ft) was in position, and a fourth (west 68ft) was still under construction (Plate 20–2).[9] Vauloué's pile-driving engine can be seen

preparing for the east 68ft centre. The date is strengthened by the unfinished air of the new block of houses to the south of Bridge Street and
along New Palace Yard, and also by the south-west tower of Westminster
Abbey not yet appearing above the roof of the nave. The tower was begun
in 1739 and finished in 1744. Evidence for a date in mid-1742 can be confirmed by the appearance of the Lord Mayor's barge. This would not
normally have been at Westminster until the state procession in October
but there were three Lord Mayors in 1742—Robert Godschall (Ironmongers' Company) who died in June, George Heathcote (Salters'
Company) who filled the office for four months only, and Robert Willi-
mott (Coopers' Company) who was sworn in as usual in October. The
painting could therefore either commemorate the mayoralty of Godschall,
a staunch anti-Walpole man who had lately been passed over by the Court
of Aldermen at no less than five elections, or represent the Lord Mayor's
procession to Westminster on 29 June 1742 when 'His Lordship in the
City Barge attended by that of the Salters' Company, went to Westminster
and was sworn in before the Barons of the Exchequer'.[10] A note in the
Salters' Company Minute Book says that the court 'proceeded in their
Gowns with their Coullers to attend the Right Honourable George
Heathcote Esq Lord Mayor to Westminster in two covered boats with two
pairs of oars for their Servants and Musick. From whence they returned
back with his Lordship to their Common Hall to Dinner at his Lordship's
request'.[11]

Unfortunately it is not possible to identify the Salters' colours or
emblems in Scott's painting but in any case the company possessed no
barge and probably hired one from another company.[12] However, the view
proved to be a popular subject and Scott painted at least six versions
between 1742 and 1750—two with the Ironmongers' arms on the sternpost,
three without arms, and one without a state barge at all. One was probably
painted for Scott's fervent admirer, Sir Edward Walpole, but the only
documented version is now in the Bank of England and was painted for
Sir Edward Lytelton in 1749. Letters from Scott to Lytelton are also in the
bank archives and relate to arrangements for delivering the painting
together with its companion to Lytelton's house. Scott asks him to confirm their safe arrival and assures him that 'they are thought to be the two
best Pictures I Ever Painted'.[13]

The Westminster backgrounds in Scott's paintings are of some interest
and qualify Horace Walpole's opinion that 'he often introduced buildings
in his pictures with consummate skill'. Indeed the Abbey in several of
his pictures reveals extraordinary distortions. Its most exquisite feature,

the Henry VII chapel, is not painted in at all; Wren's copper cupola surmounting one of the buttresses near the Chapter House is exaggerated out of all proportion; and Hawksmoor's north-west tower is so twisted out of alignment with the axis of the nave that some other explanation is needed than the distortion caused by a camera obscura. The corpus of Scott's painting shows that he was perfectly capable of understanding an architectural problem of this nature. Deliberately to misinterpret it must surely mean that it was his intention, or possibly that of a patron who admired the new tower, to give prominence to this new feature of the Westminster scene. The Abbey authorities had ordered in 1734 'that the North West Tower of the Abbey Church be forthwith raised and carried up according to the plan of Mr Hawksmoor this day produced in Chapter'. But Hawksmoor died in 1736 and the work was finished by John James who had succeeded him as surveyor of the fabric of Westminster Abbey. James drew a 'North West Prospect of Westminster Abbey with Spire as design'd by Sr Christopher Wren' and Fourdrinier's engraving of this, published in 1737, was undoubtedly used by Scott on more than one occasion.[14] This would account for the transposition of the north-west tower and also for the addition of Wren's spire on at least two of Scott's views of Westminster Bridge.[15]

Apart from these deliberate distortions, which were doubtless also intended to enhance the dramatic value of his backgrounds, Scott had no hesitation in juggling with the time sequence of his paintings, building them up from appropriate pages of his sketch-books. A view of Westminster in the Paul Mellon Collection illustrates this point clearly (Plate 24). Three distinct time phases can be seen. The first was some time during the year 1744 shortly before the completion of the south-west tower, which appears encased in scaffolding and with the stonework finished only as far as the upper course. The accounts of the Abbey carpenter, John Bacchus, show that the scaffolding was struck towards the end of 1744, therefore the painting was probably done from drawings made during the spring of that year or even late in 1743.[16] The second phase of this picture is represented by the bridge itself, which is shown as it was at some time during the autumn of 1745. The five west arches are complete and the balustrade is finished as far as the west 68ft arch, that is to say, between the completion of the middle arch balustrade (6 August 1745) and the contract for building the remainder of the balustrade over the west 56ft and 52ft arches (17 December 1745). The third phase is strikingly demonstrated by the City barge in the foreground. The morning sun shines on the escutcheon of the Ironmongers' Company decorating the sternpost. The Ironmongers'

35. Westminster Bridge from the South-East Abutment, April to May 1747, by Canaletto. (*Duke of Northumberland's collection, Alnwick. Photograph by Country Life*)

36. The damaged arches being dismantled, April to May 1749, by Canaletto. (*Royal Collection, reproduced by gracious permission of Her Majesty the Queen. Photograph by A. C. Cooper*)

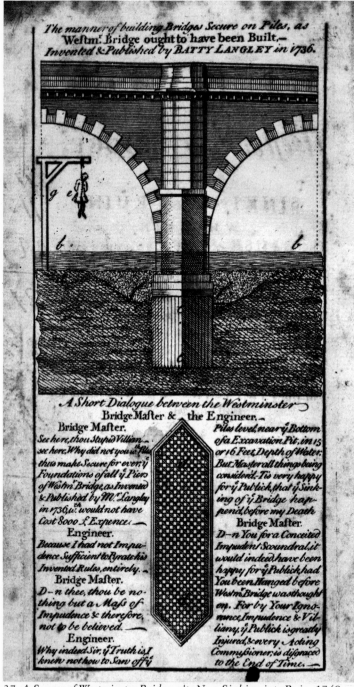

37. *A Survey of Westminster Bridge as'ts Now Sinking into Ruin*, 1748. Frontispiece to Batty Langley's pamphlet. (*Photograph by courtesy of the British Museum*)

emblems—the shackles, the three gads of steel, the helmet and terse—
are all easily identifiable in the bright sunshine, leading to the conclusion
that the bargemaster of the company was carrying out a morning rehearsal
before bringing the new Lord Mayor, Sir Samuel Pennant, to take his
oath in October 1749 with the usual pomp. Unfortunately for absolute
accuracy, the supporters seem to be a mermaid and merman instead of the
Ironmongers' salamanders. Possibly Scott used a drawing of the Fishmongers' barge from his sketch-book. [17]

Scott continued to produce paintings of Westminster Bridge long after
its completion in 1750 and probably until his departure from London to
live at Bath in about 1770. The subject was popular and made him money
and it obviously fascinated him apart from any rivalry there may have been
with Canaletto. The magnificent architectural background, the teeming
life of the river with its myriad boats, barges, ferries, hoys and shallops,
some under sail, others with oars, the riverside inhabitants, the reflections
in the water, the sky and racing clouds, all combined to appeal to the
genius of a sensitive London artist. And the loving care with which he
cherished the subtle colour combination provided by Labelye's use of
Portland and Purbeck stone shows a perceptiveness which the great
Venetian sometimes missed in his English views. The fact that neither
Canaletto nor Scott even so much as hinted at the existence of that great
evil of London life, the thick pall of smoke which hung over the rooftops
from daybreak onwards, is characteristic of the dualism inherent in the
social conditions of eighteenth-century England. Neither artist seemed to
be aware, as Hogarth was with such devastating effect, of the horrors
seething below the glittering surface of the townscape. Perhaps the
discovery of a Samuel Scott sketch-book, should such a treasure ever be
found, might prove otherwise. It would certainly provide a mine of
drawings of the riverside houses, the Abbey and Westminster Bridge itself
at various stages of building. Such a book would have been an invaluable
tool for the artist, in constant use supplying him with details for the com-
pilation of his Thames paintings. No such sketch-book is known but a few
drawings of Westminster Bridge were sold after his death, at Langford's
auction rooms in Covent Garden. [18]

An artist working in London at this time was Antonio Jolli, the
Modenese or Venetian follower of Pannini, who had been commissioned
in about 1744 to paint 'two pieces of perspective . . . . a pair of views most
sweetly painted on copper of the City of London, one of them containing
the old London Bridge'. [19] Both of them are still at Goodwood and could
conceivably be related to a mysterious picture which was sold at Christie's

in 1967 (Plates 25–7).[20] The view is from the middle of the river looking upstream and shows the bridge in the state of building towards the end of 1745. The last two piers on the Surrey side are finished but the scaffolding is still in place; timber centres are under the east 68ft and 64ft arches and the small abutment arch; the masonry for the east 64ft arch is nearly complete; and the centres for the three middle arches have been struck, but not that for the east 68ft arch. This would indicate a date between October and December 1745. The presence of a single centre under the west 68ft arch cannot easily be explained. This particular centre had been dismantled in December 1743 and there is no mention in the records of its being repositioned. It can only have been added by the artist either on his patron's instructions or perhaps simply to give interest to the composition.

Jolli's unfamiliarity with the London scene is easy to detect. Westminster Bridge itself is shown as a relatively flat affair with none of the hump-backed steepness which was one of its chief defects (see page 265). The Abbey is painted with architectural precision and with feeling for its Gothic magnificence, but St Margaret's church has been moved nearer New Palace Yard, perhaps to expose the face of the north transept, with its newly repaired rose window. And Wren's turrets at the east end of St Stephen's are capped with green copper onion domes, which contribute a bizarre air of the Orient to this otherwise placid English scene. The picture belonged to the Hon. Percy Wyndham,[21] who attributed it to Canaletto. It was engraved as such by W. M. Fellows and appeared as an illustration to Smith's *Antiquities of Westminster* (1809). But Canaletto did not arrive in England until Westminster Bridge was much further advanced than this, and moreover the painting of the buildings and the limpid translucency of the Thames atmosphere is unlike Canaletto's matter-of-fact approach.

Jolli's work in England is quite well documented, although Vertue does not mention him and Horace Walpole visited his house only to pour scorn on his ignorance of lapidary Latin. He had some sort of position at Covent Garden, probably as scene-painter and assistant manager, and painted several views of Whitehall and the Thames as far up as Richmond, where a good case has been made for his painting the decorated hall of Heidegger's House in Maids of Honour Row.[22] Indeed it is possible that the Palladian bridge in one of the panels was inspired by Westminster Bridge, which Jolli must have seen frequently on his journey's between Covent Garden and Richmond.

The artists mentioned so far—the Griffiers, Jolli and Samuel Scott—can be described as provincial men scarcely heard of beyond the English

Channel compared with the shining meteor that descended on those shores in the summer of 1746. 'Signor Canaletti (a sober man turned of 50)' was well known to English collectors, and indeed a large proportion of his output was already finding its way into the great houses of Goodwood, Woburn, Ashburnham, Holkham and Alnwick, following the visits of members of the families to Venice and Rome, where, as Lady Mary Wortley-Montagu rather harshly said, they earned themselves 'the glorious Title of Golden Asses all over Italy'. With Europe plunged into war, foreign travel became difficult. Horace Walpole and Thomas Gray managed the Grand Tour together in 1741 but after the Battle of Dettingen it virtually ceased until the Peace of Aix. Lured therefore by the prospect of earning a comfortable living in England at the expense of his noble patrons, and no doubt curious to see the splendid new bridge which Consul Smith would certainly have told him about, Canaletto arrived in London with a letter of introduction to the Duke of Richmond. Except for a brief visit to Venice in 1750 he remained in England for about ten years, finally returning home after painting the view of old Walton Bridge for its promoter, Samuel Dicker. [23]

The Thames with its vigorous life and colour made an instantaneous impact on Canaletto. He produced more than fifty paintings and drawings of London and the river from Greenwich up to Hampton Court; and of these, eight paintings and eight drawings were of Westminster Bridge itself. At the time of his arrival in London, towards the end of May 1746, the bridge still had three incomplete arches on the Surrey side, and the balustrade on the Westminster side was only half built. But of course it was a scene of tremendous activity, with hundreds of workmen swarming all over it, boatloads of sightseers sailing and rowing under its arches, Labelye in constant attendance, and visited frequently by its commissioners, many of whom were anxious to have paintings of it from the hand of the Master himself. Probably one of the first of the great series of paintings to emerge from his studio was the prospect of the Thames looking downstream and extending from St John's Smith Square to St Paul's Cathedral and with Lambeth Palace on the extreme right (Plates 28–30). It was possibly painted from the tower of St Mary's Lambeth and was bought by a Bohemian nobleman, Prince Lobkowitz, who was visiting England that summer to improve the stock of his racing stable. [24] Prince Lobkowitz—according to Walpole, 'a travelling boy of twenty . . . under the care of an apothecary and surgeon'—came to England on several occasions and was suitably fêted as befitted the son of a distinguished allied general. The catalogue of the family collection lists this view, together with its companion piece of the Lord Mayor's procession, as being bought in about 1752, though it certainly represents the bridge years earlier, in the summer of 1746.[25] The timber centres are still under the five arches on the Surrey side (they were struck in September, October and November) including the small abutment arch; the balustrade is complete over three Westminster arches and the middle arch; and work is still very much in progress at the Surrey end, with scaffolding and derricks hoisting stone from hoys below. As in all Canaletto's paintings of the subject, the steepness of the incline is clearly defined.

In the Lobkowitz painting Canaletto's interest in the accuracy of building minutiae is unusual. His most celebrated painting of the subject, 'Lord Mayor's Day', formerly belonging to the Duke of Buccleuch and now in the Mellon Collection (Plates 31–2), shows Westminster Bridge idealised to such an extent that architecturally it is almost unrecognisable, embellished with domes on all the recesses, with statues of Thames and Isis over the middle arch, and with all the timber centres removed, though in fact the last centre was not struck until five years later. The view shows the procession of the Lord Mayor, William Benn of the Fletchers' Company, to Westminster on 29 October 1746.[26]

> Last Wednesday Noon the Right Hon William Benn Esq. Lord Mayor of this City, went in great State from Guildhall to the Three Cranes, and from thence to Westminster by Water, where he took the usual Oaths of Qualification, and walked in Procession round the Hall, the Courts being all sitting, and saluted them according to custom; from thence he returned to Black Friars by Water, from whence he went in his State Coach and came to Guildhall, amidst the Acclamation of the greatest number of People ever known on the like Occasion, where a great Entertainment was provided at which were present several Lords of the Council, Nobility, Foreign Ministers, Judges and diverse other Persons of Distinction.[27]

According to Parr's engraving of the picture, published the following March by John Brindley and called 'The South-East Prospect of Westminster Bridge',[28] the barges present on this occasion were the City Barge itself, 'the finest in Europe built in the Mayoralty of Sir John Barnard'; the Stationers' barge, firing a salute to the company's patron, the Archbishop of Canterbury; the Goldsmiths' barge newly built for Sir Richard Hoare, 'with proper ornaments suitable to the Company, as handsome as the Barges of the other Companies of the City';[29] and barges belonging to the companies of the Skinners, Clothworkers, Vintners, Merchant Taylors, Mercers, Fishmongers and Drapers.

Although of course the bridge was far from complete on this ceremonial occasion, Canaletto chose to paint it as he imagined it would appear when finished (see page 168), even to the extent of surmounting the middle arch with magnificent statues of the river gods Thames and Isis (Plate 32). Drawings for these statues existed but as far as is known they were never made, though a few years later it was suggested that they would form admirable ornaments for Mylne's new Blackfriars Bridge.[30] They consisted of the arms of Westminster between the two emblematic figures and would probably have been given to Sir Henry Cheere for execution in his sculptor's yard on Westminster Bridge property. The idea appealed to Canaletto's eye for a decorated skyline, but no other artists seem to have been interested and the Bridge Journals make no mention of any decoration beyond a scroll in a stone rectangle. The idea was certainly active, however. Two years after the opening, 'one Mr Marshal presented a Petition and Plan to the Committee for making additional Ornaments to Westminster Bridge which the Committee perused and then returned to the same Mr Marshal telling him they had neither authority nor ability to proceed therein'.[31] The severe taste of Labelye and Pembroke prevailed and the bridge remained unadorned, but Canaletto continued to use the river gods unabashed, basing his work on a drawing made shortly after his arrival from Labelye's scale model, and 'improved'.

Canaletto painted views of the Thames at Greenwich, Whitehall, Chelsea, Syon House and Windsor, and journeyed further afield, probably in 1748, to Badminton, Warwick Castle and Alnwick. But he found the English countryside alien to his urban Venetian temperament and came back with relief to Westminster. The new bridge was undoubtedly a fascinating subject on which he could lavish his skill and he painted and drew it from all angles. He had no difficulty in finding a market and launched into engravings to meet the demand. The view of the City through one of the arches, painted probably early in 1747 (Plate 15), was engraved by Remegius Parr, dedicated to Sir Hugh Smithson, and advertised at three shillings and sixpence as

A Most beautiful view of the City of London, taken through one of the Centers of the Arches of the new Bridge at Westminster, and engraved from a Painting by the famous CANALETTI; *great care has been taken to have this Print well Executed*, . . . very proper for the Ornaments of Gentlemen and Ladies' Apartments, Noblemen's Halls, &c. in the Country. Care will be taken to convey this and the under-mentioned Prints into all the different great Towns in England: (*for the Curious some will be Coloured*) . . .[32]

It must represent one of the four centres on the Surrey side, removed in September to December 1746 and April 1747—possibly the last centre to be struck, 7 April 1747. A similar view but without the centre and bucket is in the Royal Collection (Plate 34).[33]

Another view painted at this time (early 1747) shows the bridge from the Stangate wharf near the Lambeth abutment, with workmen building the balustrade and hoisting stone from a barge to the east 6oft arch. This is a good representation of Westminster Bridge as it must have appeared in about May 1747, a year after the artist's arrival in England and a few weeks before the signs of settlement were detected in the faulty pier (Plate 35). The domed recesses are shown correctly, the river gods have not been added, the subtle combination of Portland and Purbeck is admirably appreciated, and the excessive height of the balustrade can be judged by the workmen standing on the south cornice.[34]

Two years later (May 1749) Canaletto made a drawing from the opposite bank showing the damaged arches being dismantled (Plate 36).[35] At about this time the strange and malicious rumour began to circulate insinuating that he was an impostor and in fact his own nephew, Bernardo Bellotto. Apart from an infinitesimal deterioration in the quality of his painting since the two magnificent London views at Goodwood, there is absolutely no stylistic evidence to confirm this story. Nevertheless it existed, and Canaletto felt the need to scotch it with an advertisement in the papers inviting doubters to attend his studio in Golden Square.[36] There are however two pictures—a painting and a drawing—which are so curious that they may have some relevance to this rumour. The drawing is in the Mellon Collection (Plate 23), the painting privately owned.[37] They are closely related and show Westminster Bridge under construction upstream on the left, the York Gate and Water-tower on the right, and various figures and boatmen in the foreground. There are other versions but these two are both undoubtedly the work of Canaletto, far beyond the range of any other artist working in London at the time. But Westminster Bridge itself is at a stage of construction, with only five arches turned, long before Canaletto's arrival in England. It is well known that great care should be taken in dating Canaletto's work. Paintings sometimes preceded drawings of the same subject, and either could be produced at any time to suit a client's whim. But in this case the view of the bridge looks as though it could only have been made with the help of a drawing or painting by another artist, probably Scott or Jolli. Canaletto could never have seen it in that condition. The fact that Scott painted several versions from the same viewpoint (Tate Gallery, Earl of Wemyss, Lord Malmesbury), using the same Thames barge moored to the York Gate steps with its mainsail furled to the masthead, makes some sort of collaboration extremely likely. There is no record of Scott and Canaletto working together but such a partnership, even in the most tenuous form, could conceivably account for the spread of this rumour. It was given a sensational twist by the watercolour painter Edward Dayes, who accused the dealers of keeping up the pretence on purpose in order to maintain the market for indifferent copies of Canaletto's work.

A painting by Canaletto showing the bridge during the repairs of 1750 was actually made four or five years later for the collector Thomas Hollis (Plate 41). The balustrade is again shown with too many domed alcoves— a puzzling fault in an artist usually so meticulous, particularly as by this time (1754) he must have been completely familiar with the details of the superstructure. The river gods have at last disappeared but work on the damaged arches is in full swing and the scene, though painted years later, was no doubt built up from notes made in June or July 1750.[38] The arches were keyed on 20 July.

Westminster Bridge, as we have seen, was completed and opened in November 1750. No painting of the actual ceremony exists. It took place at midnight but the jollifications continued throughout the next day and the Thames was thronged with rivercraft of every type. It would have made a magnificent subject. However, a painting showing the scene a few months earlier was painted by Canaletto and shows the state procession by water on 25 May 1750 when John Blachford of the Goldsmiths' Company was sworn in at Westminster, 'to which he went in the City Barge, attended only by the Goldsmiths' Barge'.[39] Two more barges near the Surrey bank were probably added for effect, and once again the surplus of domed alcoves and the statues of the river gods over the middle arch emphasise Canaletto's habit of generalisation. The procession is smaller than the usual October display and the Goldsmiths' emblems and supporters (leopards' heads and two unicorns) can dimly be discerned on the sternpost.[40]

After the opening ceremony Canaletto, Scott and others continued to paint the bridge, which—with the Abbey, Westminster Hall, St Margaret's, St John's, the House of Commons, Lambeth Palace and the usual colourful assortment of rivercraft—went on providing a popular subject for many years to come. One of the most accurate and rewarding of these views is a large and meticulously executed line engraving by Thomas Willson—'An Exact Prospect of the Magnificent Stone Bridge at Westminster with a view of the Abby [sic], Lambeth Palace and other Buildings

&c up the River Thames' (Plate 44). By means of an imaginary airy view-point above the middle of the river, the artist enables the observer to look down slightly on the roadway on which the top of a carriage can be seen and the heads of a number of pedestrians just below the level of the balustrade, confirming the complaint that the view was obstructed by a massive barrier of stone (see page 265). The steepness of the pitch is right, the domed alcoves are correctly shown surmounted by tinman Walker's iron lamps, and one of the alcoves can be seen to be enclosed for use by the watchmen. The scale of the modest stone rectangle, adorned with a swag of fruit, bears very little relation to the flamboyance of Canaletto's baroque centre-piece but accords closely with the severity of Labelye's design. The Fishmarket wharf can be seen on the right near the Bear Inn, where the opening dinner took place and the King's health was drunk to a flourish of trumpets and kettle drums.

Views by the brothers Samuel and Nathaniel Buck (1749), forming part of their panorama of the Thames from Westminster to the Tower, and by Joseph Farington, exhibited at the Royal Academy in 1792 and used to illustrate his book, *Views of Cities and Towns in England and Wales,* are both rather pedestrian affairs compared to the splendours of Scott and Canaletto. But they are both topographically exact and of great value. Marlow produced a pair of paintings which were bought by David Garrick, one showing St Paul's and the City, the other Westminster Bridge and the Houses of Parliament. Constable halted his hackney cab for an hour on the bridge in October 1834 to join the crowd of spectators gazing at the fire, which, with a noise like a cannonade and amid bright blue coruscations as of electric fire, razed the Palace of Westminster to the ground and led to its Gothic revival at the hands of Charles Barry. And finally the melancholy end of Westminster Bridge, crumbling, twisted, supported by timber crutches, its noble balustrade replaced by a wooden fence, was painted shortly before its demolition at the beginning of the Victorian age. Ansdell Smythe[41] and Henry Pether, painting the sad scene by moonlight (Plate 48), close this chapter begun so hopefully with the brave new bridge glittering in the sunlight on the canvases of Samuel Scott and Canaletto.

# NOTES TO CHAPTER 16

1. Ayloffe's paper, read to the Society of Antiquaries 25 March and 1 April 1773, was published in *Archaeologia* (1775) pp 267ff.
2. Probably by Anthony van der Wyngaerde (London Topographical Society 1882).
3. De Jongh 'The Thames and London Bridge' 1630 (Kenwood), Bol 'The Thames and Somerset House' c 1660 (Dulwich), and Hondius 'The Frozen Thames' 1677 (Museum of London).
4. *Vertue Notebooks* III p 145.
5. Formerly in the Old East India Office, now the Foreign and Commonwealth Office, Whitehall; see William Foster *Descriptive Catalogue of the Paintings Statues etc in the India Office* (1924) p 18.
6. *Vertue Notebooks* III pp 61 and 63.
7. Ibid VI p 176 and 179.
8. MS in British Museum and published as *Hogarth's Peregrination,* edited by Charles Mitchell 1952.
9. The completion certificate for the fourth centre was dated 30 June 1742.
10. *London Magazine* (1742) p 358.
11. Salters' Company Minutes Book 29 June 1742.
12. A hundred years earlier, when John Evelyn considered the Thames processions to be even more stately and magnificent than the Venetian 'Marriage to the Adriatic', the Salters had hired a barge for state occasions, arranging for 'the King's Arms and Salters' Arms which are now standing in the Hall to be taken downe and beautified and carried to furnish the Barge upon the water' (Salters' Company Minute Book 21 August 1662). On the next occasion when the Salters provided a Lord Mayor (Sir Richard Glyn in 1758) the company borrowed a barge from the Brewers. It was preceded by three boats containing a Master of Defence with eight swordsmen and many other attendants wearing the company's colours, the whole procession moving to the accompaniment of four French horns, four hautboys, two bassoons, two trumpets and a pair of kettle drums (J. Steven Watson *A History of the Salters' Company* (1963) pp 107–8).
13. Letters from Scott 20 June 1749 in the Bank of England archive. Other versions by Scott are with the Metropolitan Museum (New York), and the Sun Alliance Insurance Group (London), and in private collections in England. A drawing belongs to the S. London Art Gallery, Camberwell.
14. *Wren Society* vol XI p 16. Fourdrinier's receipt for the engraving is in the Abbey Library (WAM 46234).
15. Earl of Wemyss, Gosford House, and private collection formerly Northwick Park.
16. John Bacchus and John James's accounts 1744–5 (WAM 46454 I and G).
17. Martin Holmes 'London Scaffolding and a Brilliant Painting by Samuel Scott' *The Connoisseur* June 1964 p 85.
18. Langford *Catalogue of the Collection of Prints and Drawings of Mr Samuel Scott, painter, late of Walcott Street, Bath . . . 12th and 13th January 1773* (typescript of catalogues in the Wallace Collection library).

19. Dr Jacques *A Visit to Goodwood* (1882) p 56.
20. Christie's 7 July 1967 (100) formerly Lord Moyne's and now in a private collection.
21. Percy Wyndham, son of Sir William Wyndham, Bt. and nephew of the Earl of Egremont. He later became Percy O'Brien (see *Complete Peerage* V p 36). Constable *Canaletto* 437a.
22. Edward Croft Murray *Burlington Magazine* April-May 1941.
23. Canaletto's ten years in England have been fully described by Mrs Finberg in *Walpole Society Journal* IX and X (1923-4) and by W. G. Constable *Canaletto* (1963, revised by J. G. Links 1971).
24. *Westminster Journal* 6 September 1746. Dr Burney noticed his presence in London the year before, attending the rehearsals of the opera *L'Inconstanza Delusa* at the Haymarket.
25. Max Dvorak quoted by W. G. Constable in *Burlington Magazine* June 1923 p 279; Eduard Safarik *Canaletto's View of London* (1961) with colour reproductions of details on pp 18 and 20; Constable *Canaletto* 426.
26. Constable *Canaletto* 435.
27. *London Evening Mercury* 30 October 1746 and *Westminster Journal* 1 November 1746.
28. 'For the curious a few will be colour'd from the original Painting', *General Advertiser* 25-7 March 1746/7. The title of the engraving, 'The South-East Prospect', seems to have been a mistake, the view being of the north aspect looking due south.
29. Goldsmiths' Court of Assistants 21 February 1744/5.
30. *Gentleman's Magazine* (1754) p 60.
31. Bridge Minutes 28 November 1752. 'One Mr Marshal' might have been a descendant of the seventeenth-century sculptor Joshua Marshal, who worked in Whitehall and Westminster and left behind a considerable quantity of statuary.
32. *General Evening Post* 2-4 June 1747.
33. Constable *Canaletto* 412 and 732.
34. Ibid 434.
35. Ibid 751.
36. Finberg *Canaletto* pp 29-34; Constable *Canaletto* pp 35-8; the advertisement appeared in *The Daily Advertiser* 30 July 1751.
37. Constable *Canaletto* 427 and 747.
38. Ibid 437b.
39. *Gentleman's Magazine* (1750) p 254.
40. Earl of Strathcona's collection, reproduced in Constable *Canaletto* 436.
41. 'Westminster Bridge by Moonlight' c 1846 by Ansdell Smythe is in the Ivy Restaurant, Covent Garden.

# 17
# The Approaches

While Westminster Bridge was being built, the commissioners and their surveyors were also occupied with buying up land for the roads leading to the abutments. Improvements were desperately needed on both sides of the river. Much of the Surrey side consisted of orchards and market gardens supplying a valuable contribution to the life of London, but a great deal was marshland and the source of mists, damp and unhealthiness. At Westminster conditions were not much better. Members constantly complained of the stink in the House of Commons exuding from the river. A contemporary guide-book described Westminster as 'a kind of large forest of wild creatures, ranging about at a venture, equally savage and mutually destructive of each other.'[1] The maze of narrow streets and alleys and squalid dwellings infested with gangs of robbers made it unsafe to walk about after dark, and Justices of the Peace had been attacked even in St Margaret's churchyard, 'by divers idle persons who in the dark season lie in wait'.[2] Watchmen patrolled the streets with staff and lantern, and proclamations offered rewards of £100 and more for information leading to convictions, but both seemed to be equally ineffectual.[3] Hogarth's satirical engravings and Fielding's *Jonathan Wild* and *An Enquiry into the Causes of the Late Increase in Robbers* are horrifying enough. The reality was infinitely worse.

The first Westminster Bridge Act (9 George II, c 29, 1736) described the streets on either side of the river approaching the bridge as being 'so narrow that they may be incommodious to coaches, carts and passengers and prejudicial to commerce'. It tacitly appreciated the evil and authorised the commissioners to treat with landowners for the conveyance of as much land as they should need. Later legislation increased the commissioners' powers, and the tenth Bridge Act (1745) enabled them to supervise the architecture of the street fronts and preserve 'the Beauty Regularity and Uniformity of the said Buildings Streets Ways and Passages'. An examination of maps and plans of Westminster made between 1734 and 1761 shows how much the development of the area was due to the Bridge

Commissioners (Plates 2–3). Early in 1740 the bridge surveyor, Thomas Lediard, produced a plan of the intended new streets between New Palace Yard and the Privy Garden, and from King Street down to the river. A more detailed plan was also made showing the area from New Palace Yard northwards as far as the wharfs adjoining Richmond House, and from Parliament Street to the Westminster Quay. The quay was to have been an attractive embellishment consisting of a long open space planted with trees and stretching from the abutment steps to another flight of steps 400 feet downstream. But the ground was far too valuable for such a frivolous public amenity and it remained the gardens of the great riverside houses there—Manchester House, Dorset House, Derby House and Richmond House. It would have been a magnificent improvement. The line of Kent's new Palace of Westminster can be seen in the large plan, beginning between St Margaret's Lane and Westminster Hall (Plate 2), and if these two schemes had materialised Westminster would have become as fine a river town as any in Europe. However, the commissioners at least approved, and Lediard was ordered to have the engraving made.[4]

The commissioners' aim was to improve the tone of the areas near the bridge, especially in Westminster, and though Lediard's little quay was lost, in general their ambition was achieved. Much of this aim was Lediard's own, rooted in the 1730s in his early association with the Society of Gentlemen and the Horn Tavern meetings. His ideas were carried on by his son, who by 1745 was in the throes of negotiations with the Archbishop of Canterbury, the Dean and Chapter of Westminster Abbey, the Woolstaplers, Christ's Hospital, and innumerable private landowners. 'Upon the whole', he wrote in a report to the board, 'I am humbly of opinion that there now is a very fair opportunity to set the whole scheme in motion which has so long laid dormant, to build a large tract of ground in an elegant manner and to procure a set of substantial inhabitants, gentlemen of rank and condition, and of course encourage tradesmen to take houses of lesser rent in Bridge Street, Parliament Street, &c. and all this without any expense to the Commissioners'.[5]

Lediard junior was ably supported by the Earl of Pembroke, who struggled for years to ensure that at least the architectural decencies should be upheld on the bridge lands. In 1740 Pembroke got the board's approval for the elevation of a row of Georgian houses to be built in New Palace Yard by his close associate at Wilton, Roger Morris. But his main labours were connected with Lediard's appalling problems of acquiring land, issuing building leases to builders and speculators, and urging the board to take decisions. For instance, early in 1747 Pembroke addressed the commissioners on the subject of leases in Westminster between New Palace Yard and Whitehall, bitterly regretting that after ten years none of the ground had been let, and 'the Commissioners had not been able to avail themselves of one Farthing either by Rent or selling the Fee of the Ground which lies waste and useless'. He urged the board to come to some conclusion or else the matter would become too complicated for them to end 'the Trusts reposed in them with Honour'.

> They have never had, desire or expect any Reward for the infinite trouble and anxiety they have given themselves for above eight years for the benefit of the Publick. I do not mean their Attendance at this Board which (by the manner it has been attended) one may conclude is disagreeable enough, but I am very sensible that it is a Trifle to what those have suffered, who have in the Interims of our Meetings employ'd their Thoughts to carry on the Business of this Board in despite of the many Idle Foolish Reports & Interested Views they must have observed and must have been forc'd to overlook but not without great Vexation to themselves.

Pembroke mentioned, without actually naming him, one of the commissioners who had resigned because he had been wise enough to foresee 'that his indefatigable care was not sufficient to prevent the Board from making a confus'd and bad end' with regard to opening the approaches.[6] At a later meeting Pembroke asked that his report should be expunged from the Minutes because it had been intended only as an aide mémoire for him to speak from. The clerk, however, allowed it to stay, and it remains one of the few direct glimpses into the personal frictions and animosities which undoubtedly must have existed amongst the commissioners and their officers.[7]

Pembroke's outburst induced the commissioners to bring in an outside surveyor, James Horne, to advise on the rents they should ask for the various parcels of land listed in Pembroke's report. Horne produced his report at a board meeting in December 1747. It was dated 15 June and covered both sides of Bridge Street and both sides of Parliament Street. A plan of the area was hung up in the coffee house at the corner of Bridge Street. It was also published in the newspapers, the sites being offered at twenty-six years' purchase, the King's tax to be allowed in full, and the coachways and footways to be pitched and paved at the commissioners' expense. An officer, usually Lediard, attended regularly every morning at the coffee house. He said that most people he had treated with had asked whether they were to be tied down to any particular fronts to their houses and what special trades might be objected to.[8]

One of the trades thought to be unsuitable to this elegant neighbourhood was the fishmonger's. A petition presented to Parliament by the inhabitants of Westminster had argued that a free fishmarket in Westminster as well as at Billingsgate would bring employment to the fishermen and reduce the price of fish for the consumers of London and Westminster. The petition was successful and the Westminster Fishmarket Act was passed in 1749, the Bridge Commissioners being authorised to grant a piece of ground near the abutment.[9] The exact position of the fishmarket in relation to Westminster Bridge can be seen in Fourdrinier's plan of 1761. It occupied land on the north side of the Westminster abutment corresponding to the Bear Inn built by Andrews Jelfe on the south side.[10] A temporary market was opened in January 1750 and the main market was ready at the end of 1751.[11] But it was a failure from the beginning, mainly because of powerful opposition from Billingsgate itself, whose operators persuaded the North Sea and Channel fishermen to deposit most of their hauls well below London Bridge, even as far down as Gravesend, in order to cut out the Fishery Inspectorate. Fielding felt strongly on the subject.

> And first I humbly submit to the absolute necessity of immediately hanging all the fishmongers within the Bill of Mortality . . . for surely, if a few monopolising fishmongers could defeat that excellent scheme of the Westminster Fishmarket, to the erecting which so many Justices of the Peace, as well as other wise and learned men, did so vehemently apply themselves . . .[12]

The commissioners did their best to protect the market. James Mallors, a local property owner in Westminster and Whitehall, had put up a row of houses in the neighbourhood, one of them let to a fishmonger. Mallors was told 'that the Commissioners were much offended at his having let the house to a Fishmonger without first acquainting the Commission with his intention, as he could not but be sensible that the carrying on a Fishmonger's Trade in that place must be improper and an annoyance to the street'.[13] However, Mallors was a powerful man, and after a certain amount of negotiation with Seddon, the fishmonger (William Hannington) was allowed to lease the house and carry on selling fish there until Christmas Day 1766.[14]

By 1760 the Westminster Fishmarket was generally recognised as being a failure. The Society for the Encouragement of Arts, Manufacture and Commerce appointed Captain John Blake to manage an organisation designed to provide 'greater Plenty of Fish from distant Sea Ports & to supply the Cities of London and Westminster by Land Carriage or otherway'.[15] Captain Blake, accompanied by three of the fishmarket trustees (the Rev Dr Wilson, Sir Henry Cheere the sculptor, and Mr Ackworth) called at the Bridge Office in February 1763 to treat for a vacant strip of ground at the north side of the Westminster abutment, used as a store yard. But beyond the usual demand by the commissioners for a sight of the intended elevation no more was heard of the project and the Westminster Fishmarket ceased operations altogether.[16]

A more successful enterprise, and in fact one of the commissioners' most remarkable achievements, was the construction of Parliament Street. This was a new road driven through the slums lying between King Street and Cannon Row and opening up a broad approach from Whitehall to the Palace of Westminster and to Westminster Bridge itself. A good view of the new street, taken from the top of the Banqueting House by J. T. Smith about fifty years later, shows the dense blanket of smoke that lay over the rooftops of Westminster at this period and is mysteriously absent from the stately prospects of Canaletto and Samuel Scott.[17] A beginning had been made in 1745 when Lediard complained bitterly to the Works Committee that he was becoming 'the object of hatred of all the meaner sort of inhabitants of this part of Westminster and frequently met with insults while surveying and looking over the houses.'[18] Tenants were given notice to quit by Michaelmas 1746 and the plans, drawn up by Lediard and Labelye, proposed a street seventy feet wide with a gentle slope upwards from Bridge Street to the Plantation Office. Builders were to keep the fronts of their houses parallel and trade signs were to be hung flat over the doors and not to project outwards.[19] Jelfe and Tufnell and the paviour Thomas Phillips were ordered to pitch the road and pave the footway under Labelye's directions. Canaletto's celebrated paintings at Goodwood and Bowhill show Parliament Street in about 1747 and 1751–2 during the course of this development.[20] It was finished in 1756 and described as 'a very handsome and spacious new street, adorned with very handsome buildings. It extends from New Palace Yard to the Cockpit'.[21]

The Bridge Commissioners' work was not finished with the completion of Parliament Street. Two Acts in 1756 and 1757 authorised them to continue broadening the new street as far as Charing Cross, the sum of £10,000 being granted for the purpose. A subsidiary clause relieved James King's two sons of the family lease at the Surrey abutment.[22] The board spent some time inspecting a plan produced by their surveyor, John Simpson, showing the existing streets between the Holbein Gate and

Charing Cross, and decided that they ought to acquire all the houses on the west side of Whitehall between the Admiralty and the passage into Spring Gardens. The owners were given notice to quit and advertisements were published in the papers inviting proposals specifying terms and ground rents for building leases along the new street. Simpson produced a schedule a few weeks later valuing the property at £25,344—an amount which so frightened the commissioners that they announced it to be altogether beyond their means. In order to avoid all the cumbersome business of summoning juries and bothering individual commissioners during the summer, they seriously thought of accepting a proposal made by two building speculators, James Mallors and John Lambert, that the whole job of widening the street should be handed over to them. Mallors insisting that the £10,000 should be transferred by Act of Parliament. The two Members for Westminster, General Cornwallis and Sir John Crosse, were deputed to lay the facts before the Commons as soon as possible, together with the commissioners' opinion. Simpson took fright and submitted a revised estimate of £13,783 which was accepted. The property was bought and the new street paved by Thomas Phillips and the footway by Gayfere and Tufnell.[23] The famous Holbein Gate was taken down at last, after years of rearguard action by the preservationists, and eighteenth-century Whitehall was finished at the end of the *annus mirabilis,* 1759.[24] Legal formalities continued for years afterwards, two landowners—the Earl of Dysart and Francis Watkins—proving particularly difficult and involving the commissioners in empanelling two juries, presenting a counter-petition in Parliament, and defending their mason, Thomas Gayfere, in an action in the Court of King's Bench before Lord Mansfield.[25] John Rocque's two plans of London and Westminster, published in 1746 and 1755, show the development of Parliament Street and Whitehall; and Hogarth's engraving 'Night' (1738) gives a lurid if slightly exaggerated idea of the narrowness of the top of Whitehall before the improvement. The development of Westminster and Whitehall was criticised by John Gwynn, who accused the commissioners of throwing away a marvellous opportunity. 'The building of the new bridge and the powers with which the Commissioners were vested, demanded much more, and had a general plan of improvements been considered, it is certain that a very different use could have been made of so desirable a field for the execution of taste, elegance and magnificence'. Gwynn's own improvements were illustrated with coloured plans and were an interesting and prophetic example of enlightened town planning.[26] But his strictures were too harsh, and the commissioners' work, though obviously not so com-

38. William Stukeley's proposal for relieving the sinking pier, 1748. (*Drawing in the Bodleian Library.*
*Photograph by courtesy of Bodley's Librarian*)

39. *Thames Panorama* by Robert Griffier, 1748. (*The Duke of Buccleuch's collection, London. Photograph by A. C. Cooper*)

40. Detail of Plate 39. The balustrade of the middle arch.

prehensive in scope as the great schemes envisaged by Inigo Jones, Wren and William Kent, contributed much towards cleaning up the squalor and misery characteristic of this part of London.

On the other side of the river the commissioners' task was no less complicated. The Act of 1739 authorised them to treat with landowners 'beginning at the abutment and from thence to any part of the road leading from Church Street in the Parish of Lambeth to the Kentish Turnpike, between the said Turnpike on the left hand and the wall of the Arch-bishop's garden on the right, not exceeding the breadth of one hundred yards in any part thereof'.[27] Lediard senior's investigations into the ownership of ground needed to open a road from the bridge eastwards showed that most of the marshland and gardens belonged either to the Archbishop of Canterbury or to the City of London. Nearly all of it was let and sub-let to market gardeners and a great deal to a certain John Lawton, who had acted as Jelfe and Tufnell's security from 1738 to 1740 for the piers, abutments and three middle arches. The fee of the gardens had been leased by the Archbishop to Lawton and his heirs for twenty-one years from 1738 and after Lawton's death his widow had agreed to give up her interest for nothing. Unluckily she had sub-let part of the land to two gardeners, Gold and Kidwell, and a strip of this had been sub-let again to a Captain Pottenger. All this would have to be bought, together with another small strip between the gardens and the road called Lambeth Marsh belonging to a City man who Lediard believed would sell on reasonable terms. Gold was prepared to sell outright, but Kidwell raised objections and his interest had to be settled by a jury.[28]

This was typical of many transactions carried out by both the Lediards—some easily, others with more difficulty. The archbishop himself agreed to part with the reversionary fee for a small consideration, provided it could be arranged in such a way as to justify him to his successors—that is to say, with the sanction of a jury. Ayloffe called at Lambeth Palace per-sonally and reported that though the archbishop was extremely polite and apologised for giving the commissioners so much trouble, he was determined that a jury's verdict was essential in order to over-rule any objections that might be raised by future archbishops. Lediard was ordered to provide the archbishop's steward, James Parry, with a plan of the ground needed, a jury was empanelled and eventually the difficulty was amicably settled.[29]

Though the archbishop was as co-operative as possible, Lediard junior found it a great deal easier to treat with the Common Council of the City. For instance a strip of land was wanted between the east end of the Willow

Walk at Lambeth and the sign of the Dog and Duck in St George's Fields. A memorial was referred to the committee for letting the Bridgehouse lands. They measured the ground with Lediard and found that one part had been let to two spinsters, Elizabeth and Anne Symonds, for twenty-one years from 1731. The other part, consisting of the Dog and Duck itself and about sixty-three acres of land in St George's Fields, belonged to a haberdasher, Richard Hall, and had been sub-let to a market gardener, Thomas Clarke, for sixty years from 1713. Lediard attended a meeting of the Grand Committee of the City Lands at the Guildhall and agreed on £150 for the absolute purchase of the two pieces. The question was settled in a few weeks, though a slight change in the direction of the road was decided on because of a particularly swampy patch belonging to the archbishop. In order to avoid this patch, Ayloffe sent a request to the City asking for an alteration in the sale. This was rejected at first, perhaps because the new direction had not been properly explained, but it was agreed later and the two plots were sold to the Bridge Commissioners for £150.[30]

In 1744 both Lediard and Labelye independently surveyed the ground over the meadows and reported their ideas on the best way of making the new road from the Surrey abutment to the Lambeth Marsh road. Lediard's plan for an ambitious causeway, 100 feet wide and raised twelve feet high on a series of brick arches, was rejected—not altogether surprisingly—because of the cost, the time factor and the depressing fact that nowhere in England were the roads anything like 100 feet wide. Most roads were in fact no more than twenty feet wide.

Labelye's ideas were set out in a group of reports in 1744 and 1745 and throw a light on the methods of road-building in the eighteenth century. Firstly he declared that not enough land had been bought to make the road fifty feet wide, let alone 100. The new road should be divided into three parts: a thirty-six-foot carriage-way, a single eleven-foot footway raised several inches above the carriage-way, and a ditch or kennel three feet wide between the two. The materials should consist of the old timber and bridgings of the eight centres of Westminster Bridge, a quantity of planks of oak or elm, the hardest chalk and any amount of ballast and gravel. Then, as soon as the commissioners could take possession of the ground, the new road should be staked out, raised with earth and rubbish mixed with river mud, and matched with the slope on the Westminster side. While this was going on and the brick arches over the ditches were being turned, the garden ground 'should be smoothed and hardened by degrees by rolling it over first with a large Wooden Roller (such as are now used to roll grass plats), next with a Stone Roller, and lastly with the heaviest Iron Roller that can be had, employing horses for this last'. A two-foot bed of the toughest chalk should be laid 'by hand and not thrown at random on the ground'. Six to twelve inches of screened ballast should be laid over the chalk and rained on more than once. 'The water in passing through will precipitate all the Loom, Earth, Sand and other small bodies between the interstices of the chalk and greatly add to its strength and firmness.' More gravel was then gradually laid on top and rolled after a shower of rain: 'the largest sort for the Carriage Way must be laid swelling in the middle as in the Turnpike Roads, and men must be employed at least for some time to attend to it daily, fill the Rutts when too deep and to throw on more Ballast and Gravell as often as it shall be necessary'. The materials were to be kept in place by means of stakes and planks from timber already used for the bridge. Drainage was to be managed by the kennel between the footway and carriage-way, but in addition, 'wherever Nature shall point out a necessary place, little cross drains be turned over the Chalk to carry water into the Ditch'.[31]

The report was dated 7 October 1744 and Labelye ended it with a recommendation that whatever the method decided on the materials should be ordered immediately, so that work could begin early in the spring. He produced a number of models of the road to illustrate his points but he could not persuade the board to come to a definite decision. On 23 July a meeting was held at the Bridge Office to discuss the convey-ances detailed by Lediard. On this day Prince Charles Edward landed in Scotland, and though the news did not arrive in London till early in August, no meetings of the Bridge Commissioners took place until 8 October. Meetings then continued at weekly or fortnightly intervals as usual, the prince turning back from Derby on 6 December. Work on the bridge itself does not appear to have been noticeably effected, though the direction of the road was still not decided. Finally it was staked out for the last time early in 1746 and in June Labelye was ordered to begin work as soon as 'a sufficient quantity of ballast and wooden material shall have been provided'.[32] Labelye had already found, a few miles up river, a large quantity of green wood—elm, oak, thorn, maple, hornbeam, apple, pear, plum, crab and chestnut—which could be had for a reasonable price ready cut into lengths of between five and ten foot. There was also enough yew, holly and evergreen oak to be used for the swampy places near the ditches. Henry Monk, a gardener of Fulham and the owner of the wood, agreed to deliver it bundled into fascines at the wharfside.[33]

That summer the commissioners took possession of all the land they needed. After a final survey, during which Labelye pronounced the ground to consist of 'a rich black mould extremely good for gardens & extremely bad for a road', the work was put out to tender. Robert Smith and William Godfrey were commissioned to build it and by October 1746 the whole new road was raised high enough and sufficiently levelled to receive its upper coat of bavins and Thames ballast.[34] Most of it was ready for traffic by the summer of 1747, posts and rails ('primed twice in Oyl at as cheap a rate as possible') being set up in June. On 14 November ended 'the Stowing of Thames Ballast on the New Road which is now fit for service and the publick part of it daily used'. One carpenter was retained as a labourer to fill up ruts and look after fences, posts, rails, bavins, and all drains and sewers.[35]

The new road is shown in Rocque's plan published in October 1746 (Plate 3). At first it merely opened up a fresh approach to Lambeth, Spring Gardens and the Southwark bank of the river towards London Bridge. It went for about half a mile into Lambeth, as far as the Dog and Duck, an inn near London Stone and a favourite refuge for the footpads of St George's Fields.[36] But a year after the bridge was opened, the Surrey Turnpike trustees petitioned Parliament for a network of new roads to Kennington and Newington connecting Westminster direct to the Dover Road.[37] They were opened by the king himself travelling on 31 March 1752 to Harwich via Westminster Bridge and London Bridge, and are shown in the new edition of Rocque's map, published 7 January 1755, 'with all the New Roads that have been made on account of Westminster Bridge and the New Buildings and Alterations to the present year MDCCLV'.[38]

### NOTES TO CHAPTER 17

1. *A Trip from St James's to the Royal-Exchange* (1744).
2. Westminster Court of Burgesses 6 September 1733.
3. *London Magazine* (1744) pp 464 and 568.
4. Bridge Minutes 10 March 1739/40.
5. Ibid 23 April 1745.
6. Ibid 10 February 1746/7.
7. Ibid 7 April 1747.
8. Ibid 1 December 1747.
9. Acts 22 George II, c 29 (1749) and 29 George II, c 39 (1756).
10. Middlesex Land Register 1747/3/503.
11. *General Advertiser* 16 January 1750 and 27 December 1751.
12. Fielding *A Voyage to Lisbon* 1754 (Everyman p 264).

13. Bridge Minutes 27 February 1752/3.
14. Ibid 17 April 1753.
15. Ibid 15 December 1761.
16. Ibid 13 February 1763.
17. Reproduced in J. T. Smith *Antiquities of Westminster* (1807) Additional Plates 16.
18. Works Minutes 11 June 1745.
19. Bridge Minutes 14 and 28 April and 22 December 1747; trade signs rattling in the wind were a great annoyance.
20. John Hayes 'Parliament Street and Canaletto's Views of Whitehall' in *Burlington Magazine* October 1958 pp 341–9.
21. R. & J. Dodsley *London and its Environs* (1761) V p 111; for the development of Parliament Street see LCC *Survey* X (1926).
22. Acts 29 George II, c 38 (1756) and 30 George II, c 34 (1757).
23. Contracts in Bridge Minutes 16 August 1758.
24. For Whitehall and the Charing Cross improvements see LCC *Survey* XVI where the Westminster Bridge Commissioner's efforts are given full recognition.
25. Bridge Minutes 7 December 1761 and 1 July 1762.
26. John Gwynn *London and Westminster Improved* (1766) p 10.
27. Act 12 George II, c 33 (1739).
28. Works Minutes 13 October 1742.
29. Works Minutes 27 September and 13 October 1742; Bridge Minutes 12 January, 2 and 30 March 1742/3.
30. Court of Common Council Journal June and July 1745 and 24 April 1746.
31. Works Minutes 7 November 1744.
32. Bridge Minutes 17 June 1746.
33. Ibid 7 January 1745/6.
34. Works Minutes 7 October 1746; Bridge Contracts (81) 5 August 1746.
35. Bridge Minutes 17 November 1747.
36. The Dog and Duck also possessed a medicinal spring which bubbled fresh from the garden and contrasted favourably with the stale product of Chelsea Waterworks.
37. Act 24 George II, c 58 (1751).
38. *Gentleman's Magazine* (May 1753) pp 207–9; Darlington and Howgego *Printed Maps of London circa 1553–1850* (1964) p 100.

# 18
# Decline and Fall

Grandiose and rather boastful claims were made on behalf of Westminster Bridge after its completion in 1750. A very minor poet imagined the river gods of Rhine, Danube, Tagus and Seine visiting the Thames and being thunderstruck at its beauties:

> Now mingling spires, and Paul's stupendous dome
> Attract their eyes, as westward on they roam;
> Till winding to the left, as leads the flood,
> Sprung the last wonder, and before them stood.
> Astonish'd! ravish'd! 'No confusion's here,
> The uncumber'd structure swells distinct and clear',
> They cry'd—'but whence? how rais'd? O Thames impart.[1]

The author of *Gephyralogia* was so impressed with these lines, 'composed by a friend of mine accustomed to the art', that he published the whole poem as an embellishment to his pamphlet. Describing its many virtues— its length, regularity, beauty of workmanship, spaciousness, easiness of access, and the great inland waterway which it crossed without impeding navigation—he decided that no bridge mentioned in history could equal that of Westminster.[2] The new edition of Maitland's *History of London* described it as 'one of the finest in the world . . . built in a neat and elegant taste and with such simplicity and grandeur that whether viewed from the water or by the passenger who walks over it, it fills the mind with agreable surprise . . . There is not perhaps another in the whole world that can be compared to it'.[3] The new edition of Stow's *Survey* did not describe the bridge, though it says that Bridge Street 'hath a fine prospect both of the Bridge and the Park, and is broad enough in the middle for half a dozen coaches to pass.'[4]

Labelye himself, never noticeable for understatement, claimed that it was 'certainly one of the most magnificent Bridges in the whole world . . . This Bridge has certainly nothing of the kind in Europe and perhaps in the whole world, especially if its situation and construction are considered . . .

and not only a beautiful, but a most lasting, useful and necessary communication between the neighbouring Counties and a very great ornament to the Capital of the British Empire.'[5]

In many respects these claims could be justified. The celebrated bridges in France were not built over tidal waters and did not possess the classical simplicity of Westminster. The Pont du St Esprit over the Rhône, built in the thirteenth century, was looked on as inferior because of the inequality of its arches and the angle it made in the middle to resist the flow of the river. It was also extremely narrow, being hardly wider than the bridge at Avignon, which was about twelve feet across. On the other hand it had twenty-five enormous arches, nineteen of which spanned between seventy-eight and 108 feet. The Pont Guillotière, the great medieval bridge at Lyons, consisted of twenty unequal stone arches and was also angled in the middle. It had been built narrower originally, but later a parallel bridge was put up alongside and the two clamped together. Both the Pont Royal and the Pont Neuf in Paris were five feet broader but considerably shorter than Westminster Bridge; and the great medieval bridges at Ratisbon, Dresden and Madrid were comparable. Even that choleric Philistine, Smollett, enduring the discomforts of inn life at Nice, remembered Westminster with pride and nostalgia. 'I have not seen any bridge in France or Italy', he wrote, 'compared to that in Westminster, either in beauty, magnificence or solidity; and when the bridge at Black Friars is finished, it will be such a monument of architecture as the world cannot parallel'.[6] Furthermore, as the paeans of praise by Canaletto make clear, much of the charm of Westminster Bridge lay in Labelye's skilful use of Portland and Purbeck stone in the arches and spandrels. With the sun shining on the Thames and the white and pale green bridge reflected in the water, with the Abbey and St Margaret's on one side and Lambeth Palace on the other, Westminster must have been a very attractive sight indeed.

In fact it is extraordinary that, except for various articles in newspapers and magazines, there was very little contemporary comment on the beauties of this fine new adornment to the London scene. One of its attractions, we are told, was a curious echo in the arches, and a favourite river pastime was to float underneath playing a French horn. Edward Moore, the editor of *World*, a magazine satirizing the follies of fashionable society, describes a voyage from Vauxhall to Whitehall on a dark night under a tilt.[7]

All that I shall inform my readers of this voyage is that it appears from the journal of it, kept by one of the passengers, to have been a very indiscreet one; and that in the latitude of Westminster Bridge, Miss Kitty, a young country beauty of eighteen, was heard to say with great quickness to a colonel of the guards who sat next to her, 'Be quiet Sir!' and to accompany her words with a smart slap on the face so that the centre arch rang again; upon which her aunt, who was one of the party, took occasion to observe that her niece would always be a country girl and know nothing of the world.

Another feature was the echo in the alcoves. This was described by Gayfere himself to the sculptor Nollekens. 'If you go to one of the middle alcoves and speak in a whisper, putting your mouth close to the wall, to a friend on the opposite side, after he has placed his ear close to the centre of the other alcove, he will hear every syllable you utter as distinctly as he would if you had both been in the gallery of St Paul's.' Gayfere's story is hard to believe but it is confirmed by a Londoner in 1904 remembering his father's voice from the opposite alcove clearly sounding over the noise of traffic.[8]

James Boswell, with less innocence and 'in armour complete', found an alcove convenient for one of his repellent amours. Horace Walpole could only find time to express irritation at the crush of traffic on the way to Vauxhall Gardens. Even Wordsworth, in his breath-taking sonnet composed on Westminster Bridge, was inspired more by the city sleeping early on a July morning than by the bridge itself. And Dorothy Wordsworth describes how the Dover coach left Charing Cross and 'the City, St Paul's, with the river and a multitude of little boats, made a most beautiful sight as we crossed Westminster Bridge'.

Perhaps Londoners felt inhibited about their bridges. It was left to Canaletto primarily to immortalise the splendours of Westminster Bridge and Walton Bridge; and to Piranesi the intricacies of Blackfriars Bridge. Descriptions of life on the Thames are more vivid when written by foreign travellers than by the local English. 'The ingenious and learned Monsieur Grosley', visiting England in 1765, described his arrival in London.[9]

I arrived in London towards the close of day. Though the sun was still above the horizon, the lamps were already lit on Westminster Bridge and the roads and streets leading to it. These streets are broad, regular and lined with high houses, forming the most beautiful quarter of London. The river covered with boats of different sizes, the road, the Bridge and the Streets lined with coaches, their broad foot-paths crowded with people, offered to my eye such a sight as Paris would present if I were to enter it by the finest streets of the Fauxbourg St Germain or the Place Vendôme . . .

41. Canaletto. The arches being rebuilt, June to July 1750. (*Lord Hambleden's collection, Hambleden Manor. Photograph by courtesy of the Royal Academy of Arts*)

42. *An Arch at the Westminster End in about 1750* by Samuel Scott. (*Paul Mellon Collection. Photograph by courtesy of the Paul Mellon Foundation*)

43. Westminster Bridge after the removal of the centres in June 1751, by Canaletto. (*Royal Collection, reproduced by gracious permission of Her Majesty the Queen. Photograph by A. C. Cooper*)

44. *An Exact Prospect of the Magnificent Stone Bridge at Westminster, 1747. (Engraving by Thomas Willson in the House of Commons. Photograph by courtesy of the Department of the Environment. Crown Copyright*)

45. *View on Westminster Bridge 1790.* (*Aquatint by Thomas Malton from* The Picturesque Tour. *Photograph by courtesy of the Department of the Environment, Crown Copyright*)

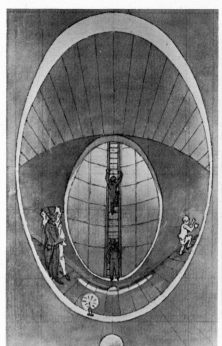

THE WESTMINSTER WATCHMAN,
GUARDING the PEOPLE'S PROPERTY.

46. Inside the hollow pier, 1796. (*Watercolour drawing by R. B. Schnebbelie in the Archives Department, Westminster City Libraries. Photograph by R. B. Fleming*)

47. *The Westminster Watchman*, 1798. Charles James Fox guarding the people's property. (*Coloured etching by Robert Dighton in the Archives Department, Westminster City Libraries. Photograph by R. B. Fleming*)

48. *Westminster by Moonlight*, 1855, by Henry Pether. (*Author's collection, London. Photograph by A. C. Cooper*)

Carl Philip Moritz, travelling in England a few year after Grosley, was also deeply impressed, at least at the beginning of his tour. He found the roads between Greenwich and London alive and full of people on foot, on horseback and in carriages, and was specially struck by the number of young healthy people wearing spectacles.[10]

> Finally [says Moritz] we arrived at the magnificent Bridge of Westminster. To cross over this bridge is in itself like making a miniature journey, so varied are the sights. In contrast with the round modern majestic cathedral of St Paul's on the right, there rises on the left the long medieval pile of Westminster Abbey with its enormous pointed roof. Down the Thames on the right can be seen Blackfriars Bridge, hardly less lovely than its neighbour upon which we ride. On the left bank of the Thames, beautiful with trees, stand terraces and among them the newly erected Adelphi Building. On the Thames itself pass back and forth a great swarm of little boats, each with a single mast and sail, in which people of all classes can be ferried across; and so the river is nearly as busy as a London street.

There was plenty of criticism of Westminster Bridge, however, apart from the storms which arose because of the sinking pier. One of its main disadvantages was the steepness of the ascent from the abutments to the crown of the middle arch. Horses found the stone paving treacherous and slippery, especially in winter, and even though the road was forty feet wide, accidents were fairly common. Bridges should be as level as possible and Westminster Bridge, in spite of the claims made for its modern look, was really only a stream-lined version of the medieval hump-backed bridges prevalent throughout Europe. Moreover the chance of achieving a splendid view and adding to the pleasures of walking in eighteenth-century London was lost. 'The length of Westminster Bridge is equal to about two fifths of the Mall in St James's Park', said a contemporary architectural historian. 'How delightful it would be to have had an uninterrupted vista this whole extent.'[11] Furthermore the view was also obstructed laterally up and down the river. The balustrades were massive and immensely high and only by riding on the top of a coach could one see comfortably over the rail. Grosley complained bitterly of this defect, which—combined with the built-up river banks—made the Thames practically invisible unless the traveller was either in a boat or looking from the window of a warehouse. London Bridge, Blackfriars Bridge and Westminster Bridge had no prospect of the river

> except through a balustrade of stone with a rail of modillions three feet high, very massy and fastened closely to each other, the whole terminated by a very heavy cornice and forming a pile of building about ten feet high.[12]

Grosley had been told that the balustrades had been purposely built high in order to counteract the natural bent of the English, and particularly the Londoners, to suicide. This seems to have been a popular belief amongst foreign travellers to England and a visitor earlier in the century wrote about it in a letter home, expressing his surprise at the light-hearted way the English committed suicide. He attributed it to the black melancholy which assaults the native and which he actually experienced himself. He was saved by a friend finding him new lodgings in Islington where he was cured by the change of air and milk fresh from the cow every morning. 'I am certain', he said, 'that most Englishmen who put an end to their days are attacked by this terrible malady of the mind, for it is very frequent in London.'[13] Certainly the newspapers of the day were full of advertisements for remedies such as Mrs Bell's famous drops for Hypochondriac Melancholy, and early in the nineteenth century, when conditions in Westminster were supposed to be improving, the Royal Humane Society protested to the commissioners about the 'numerous cases of suicide that occur at Westminster Bridge'.[14]

However Grosley's description of the balustrade is misleading. The exterior elevation of the superstructure may have formed a 'pile of building ten feet high', but the actual interior height from pavement to rail was six feet nine inches. Furthermore we read in *Low Life* of 'hundreds of people, mostly women and children, walking backwards and forwards on Westminster Bridge for the benefit of the air, looking at the boats going up and down the river and sitting on the resting benches'.[15] The relative height of the balustrade can be judged from Samuel Scott's painting in the Paul Mellon Collection (Plate 42).

Another defect was the quantities of dust raised by the traffic, especially in summer. A succession of carters were employed removing dirt and horse manure and watering the bridge in dry weather.[16] The whole surface was repaved by John Johnson in 1794 and again in 1796, the cost being rather higher because of the slate and stone tax levied in 1794.[17] But the dust was often a nuisance. In 1805, for instance, the inhabitants of Westminster and Lambeth complained

> of the very great inconvenience we daily experience in consequence of the Bridge not being properly and regularly watered, a circumstance that is attended with great injury to the property and also a great nuisance to the Publick who at times can scarcely pass over the Bridge from the Torrents of dust that accumulate thereon.'[18]

Eventually in 1823 McAdam was called in and the surface was made into a stone road, mostly at night during March and April 1824, 'with the least possible inconvenience and interruption to the public'. [19]

These were relatively minor drawbacks, causing inconvenience rather than actual danger. Labelye's basic miscalculation, which had caused the pier to sink in 1747 and which was to make Westminster Bridge eventually unsafe, lay in his method of planting the foundations of the piers on gratings on the river bed and dispensing with the usual sub-foundation of timber piles driven into the ground. So many expert witnesses, including John and George Rennie, Telford and James Walker, have given this as being the cause of its downfall, that there is no need to look further.

The trouble started again shortly after the opening. John Simpson, the carpenter succeeding Ruby, was ordered to 'survey and examine the Bridge and its appurtenances at convenient times and to report his opinion of the condition thereof with respect to any kind of Repairs that may be wanting'. [20] Simpson reported every six months or so, mentioning the condition of the road and pavements both on the bridge and in the new streets round about, any cracks in the cornice and balustrade in need of pointing, any disturbance of the river bed round the piers, and in fact keeping a weather eye on the state of the bridge in general. Since Lediard's discharge in 1750 he had also acted as surveyor, measuring and making plans of property to be leased or sold on the commissioners' behalf. At first his duties were menial. For instance in 1753 he reported that there was 'great want of proper places to piss in at the four corners of the Bridge. For the great numbers that piss there cause it to run down to the houses which make it very offensive'. The commissioners ordered him to cope with the situation and arrange for

> four stone pissing Basons to be made and fitted up with proper stone brackets and iron rims and spikes on the tops and copper plates to the sink holes, and that he also cause one of the said Basons to be fixed at each of the four ends or Corners of Westminster Bridge . . . and that the whole expence shall not exceed £32.4.0.

The basins were fixed at the corners a few months later and the fact was recorded in Simpson's report. [21] Gradually, however, he began to improve his status, and by 1754 he felt able to present a petition to the commissioners drawing attention to his increased responsibilities. He claimed to have acted as surveyor since Lediard had gone, his duties being not only to examine and report on the state of the bridge but also to act as agent for the commissioners in their affairs relating to the bridge lands. He supervised the maintenance of the bridge, and the upkeep of the streets and pavements at the approaches, and attended every meeting in order to receive the board's instructions. He found that he was able to devote less time to his own business as a carpenter and he believed that £50 a year would 'not be thought an immoderate Allowance for his skill, time and trouble'.

The board accepted the petition and Simpson was ordered to pay Simpson £50 for a year's allowance as surveyor to the commission.[22] The allowance continued annually until his death in 1774, Simpson acting as trustee together with Seddon in most matters relating to bridge property in addition to his maintenance work on the bridge itself. Most of his reports state that 'the Bridge stands well', occasionally that the abutments and balustrade needed repair, and more often that the road surface and footways were in poor condition and should be repaved. There was never any mention of the foundations being insecure.

However in 1759 rumours were going about that Westminster Bridge was becoming unsafe. The widening of the central arch of London Bridge, finished in July 1759,[23] increased the flow of the river and in spite of Simpson's optimistic reports the commissioners decided to ask four distinguished architects to give their views. Their report was handed in at the end of the year.

3rd December 1759—In pursuance of an Order from the Honble Commissioners of Westminster Bridge of the 20th November last, signified to us by Mr Sam Seddon

We whose names are under written, have viewed and examined the State and Condition of Westminster Bridge and its appurtenances in all parts with the utmost Care and Exactness: and find the same Defective in the following particulars (that is to say) The Works in some parts are rack'd and settled; the Caps of the salient Angles and the Springing Courses of the Arches are some of them out of a level, there appear Chasms in the Fascades, there are many Spalts and Fractures in several of the Archstones; the Pier which was partly rebuilt is sunk below the adjoining ones, on the South side five inches and a quarter, and on the North side three inches three quarters. The Ballustrades are in a serpentine form; the Paving is very much decayed and lies in a disagreable and irregular manner;

And we are of opinion that the properest method which can be taken to repair and rectify these defects, and to remove the bad effect which their Appearance must necessarily have with Persons of Judgement, will be to pull down and reset the Ballustrades in regular and right lines: to take up the Pavement over the centre Arch and two adjoining Piers, and raise it three inches higher: to take up the rest of the Footways and repave them with Moorstone in an inclin'd line to the extremities of the Bridge: to take out so much of the Stone Work where the Chasms are, as will be necessary (by putting in other stones) to close the same: and to repair the Spalts and Fractures in the several Arch Stones.

Geo: Dance     Jn⁰. Phillips     Henry Keene     Robᵗ. Taylor

This formidable report was qualified a week later by the rider that none of the defects mentioned could be said 'to affect the strength of West-minster Bridge so as to render the condition of it dangerous'.[24]

The four architects were paid £22 17s od for their trouble but only minor routine repairs were carried out. Cracks in the cornice and balu-strade were pointed. The benches were painted and mended every so often. Simpson's soothing reports continued to appear at six-monthly and then yearly intervals. Dr Stukeley, by this time an aged man, read another paper to the Royal Society repeating his claim to have rescued the sinking pier, 'which remains at present as firm as any other part of that admirable and beautiful bridge'.[25] Then a year later an anonymous but prescient letter appeared in the *Gentleman's Magazine* warning the public of impending danger:

> Westminster Bridge is perhaps the most majestic pile of its kind in Europe but although it appears strong yet on a critical examination it is demon-strably feeble. It is top heavy and too narrow for its height; the piers are by no means proportioned to the weight they sustain, nor do they take suffi-cient hold of the bed of the river but stand loosely on the bottoms of the caissons they are built in . . . and though the sinking may be at present imperceptible, even by a plummet, yet the immense weight of the super-structure and the sandy footing of the piers, will in time produce a very disagreable effect.[26]

For the time being the commissioners decided that the bridge was perfectly safe and there was no cause for alarm. In 1766 rumours began to circulate again and the commissioners published a statement reassuring the public that the bridge was in no danger and that none of the piers had settled since 1753.[27] However the road was completely repaved a year later by Jonathan Raine, who won the contract against William Jelfe, Gayfere and John Herobin.

Regular reports continued to be issued by subsequent surveyors (see Appendix, page 297–8), usually asking for the commissioners' approval for repairs to the superstructure, the water-stairs, the benches, lampholders and drains, but never alluding to any structural defects in the bridge or its foundations. Occasionally the piers had to be pointed where the water had washed out the Dutch tarras, and once the balustrade and coping of the middle arch was damaged by the mast of a Maidstone hoy which was 'by the strength of the tide and stress of weather forced in spite of all their endeavours to prevent it, against the Bridge'. William Jewitt, the surveyor succeeding Simpson, carried on the reports until his death in office in 1801, the only event of interest being the opening of the hollow pier in 1795–6 (Plate 46). 'This measure', said Jewitt, 'is strongly recommended by Mr Gayfere the Mason who was at the building of the Bridge.' The pier was duly opened, the counter-arches were examined and the commissioners were told that 'both arch and pier are firm and in good order . . . and we have no doubt that this hollow pier is as strong as any others'.[28]

Jewitt was followed as surveyor by Charles Craig, who had been second officer in the Office of Works since 1793. Craig continued the same optimistic reports till his own retirement in 1829, although in the meantime the commissioners decided, for the first time, that they ought to take advice from a professional civil engineer. There could be no man more capable of delivering a sound judgement than John Rennie, at that time drawing up his plans for Waterloo Bridge.[29] Rennie reported in June 1811 that as far as the boisterous weather would allow he had examined the piers and cutwaters of Westminster Bridge and thought they were perfectly safe for another five or six years, though repairs should not be delayed for too long. He criticised the shape of the cutwaters: 'I have no hesitation in saying that pointed cutwaters whose sides are straight lines such as Westminster Bridge are nearly the worst form that can be adopted'; but he admitted that not much could be done about this beyond chiselling the decayed stone as far as possible into the shape of a Gothic arch. The mortar inside the piers he pronounced to be 'of the worst kind and crumbles to dust like a heap of earth'.[30] Rennie died in 1821 but his suggestions for repair work were carried out for many years and were confirmed by the engineer of London Bridge, James Hollingsworth, who surveyed Westminster Bridge in 1825 and again in 1827.[31]

In 1823 Telford was called in to help and in a remarkable document addressed to the commissioners he analysed the trouble and suggested another remedy. Using Labelye's pamphlets, Gayfere's manuscript *Account Book* and his own soundings of the river, he was able to compare the state of the foundations after the lapse of seventy years. Where Labelye's platforms had been laid between five and fourteen feet below the river-bed, Telford found them to be no more than three to seven feet, the lowering of the bed being caused probably by the alterations to London Bridge in 1758–62. Should the whole of London Bridge be removed (which in fact happened when it was rebuilt to Rennie's design a few years later) Telford predicted that land floods and the heavy tides forced up river by storms in the North Sea would undermine them still more. His remedy consisted in driving a row of strong main and sheet piles round each pier and in paving the whole river bed between rows of piles driven across the river both below and above the bridge.[32]

No action was taken on Telford's advice until 1829. He was called in again then and reported that though little change had taken place the foundations were in a precarious state, 'the wooden platforms only now remain below the surface of the river bed'. The commissioners at last gave orders for three of the piers to be rebuilt (with Bramley Fall stone) and surrounded with a strong wall of piles. The Clerk of Works, William Swinburne, reported the following year that the work had been done on one pier but he also said that bad weather had caused a good deal of damage, particularly to the abutments. The main defect, however, still lay in the foundations. They had been undermined even more and the surrounding gravel washed away, partly by the alterations going on at London Bridge and partly by dredging operations carried on by gravel extractors working too near the piers. Acting on Telford's advice the board wrote to the Lord Mayor demanding that the water bailiffs should forbid gravel engines to work between the Lambeth Horseferry and Whitehall Stairs. Telford's last report (12 September 1831) repeated his warning about dredging too near and emphasised that because of

the very imperfect quality of the old masonry every symptom is alarming . . . the whole bridge is in so defective a state that there is no previous estimating the time or expense that will be required to restore the several parts.[33]

After Telford's death in 1834, William Cubitt was called in to advise first on what effect the proposed embankment along the Speaker's garden would have on the foundations of Westminster Bridge, and secondly on the best and most economical way of securing the foundations whether the embankment was built or not. Cubitt drew the commissioners' attention to Labelye's imperfect system of laying the foundations and repeated Telford's advice that the river-bed should be paved under the arches and up and down stream for at least fifty feet. This would cost between £120,000 and £150,000. He also suggested that the superstructure should be removed and Labelye's heavy balustrade replaced with a closed parapet so that the view of the new Houses of Parliament should be uninterrupted. This would cost another £5,000 to £10,000.[34]

A select committee [35] was then appointed to consider the various reports, coming out in strong terms against the old bridge, which, they said, had been 'originally constructed in a very inefficient manner even for the then state of the river', and in a few more years could be 'in the utmost peril'. The superstructure was 'loose, bulged and intied and as bad a piece of stone work as any in the Kingdom'. They suggested that 'it seems matter for important consideration whether the present structure should not be pulled down and a new Bridge built in its place'. This was the first occasion that a new Westminster Bridge was formally proposed. [36] Sir Robert Smirke and James Walker were then appointed to survey the bridge and advise on the best policy. Cubitt's plan was turned down and Walker's adopted—to encase all the piers in coffer-dams and completely rebuild them on piled foundations. This would cost £68,800. [37] The rejection of Cubitt's scheme to pave the river bottom drew sportive comment from *The Times'* leader-writer, who was unaware of the idea's origin with the distinguished Telford. 'The rampant energies of our paviours', he said with heavy sarcasm, 'might thus probably have been diverted into a more convenient channel, and who knows but we might have been enjoying the blessings of an open thoroughfare at the present hour.' [38]

James Walker's report ended with the opinion that financial considerations probably favoured repairing the old bridge. The sale of Bridge estates in Westminster and Whitehall would balance the cost. But this ought to be completely outweighed by the central position and importance of Westminster. 'The ornament, the greater permanency, the freedom from future repair, and the accommodation given by a wider and more level bridge' would all tell in favour of building a new bridge if money could be found.

In fact the need for a new bridge was emphasised again and again throughout the country, but year after year the patching and bolstering continued in fruitless attempts to keep the aged carcass open to traffic. Walker and Burgess accepted the situation and devoted the next ten years to the depressing job of dismantling the old bridge section by section and then building it up again. In 1841 Walker handed in an account of his firm's work over the past five years to the chairman of the committee supervising repairs, Lord Lowther. After describing briefly the building methods adopted by Labelye, the repairs recommended by Telford and others, and the work actually carried out by his firm, including the removal of Stukeley's counter-arches, [39] he ended on a hopeful note:

These works have caused us much trouble and anxiety and some extra superintendence, but they have all been successful and I believe the Bridge, so far as we have gone, to be better than on the day it was finished.[40]

During the course of 1843 the roadway was lowered several feet and about 2,000 tons of superstructure were removed and replaced with brick arches. It was only actually closed for two or three months during the winter, the traffic being temporarily diverted over Waterloo Bridge. 'Few are aware', said The Times, 'that within that huge and ugly mass of piling one of the most difficult and dangerous operations of engineering— namely that of restoring the foundations of an arch injured by the action of the tide—is at this moment in the course of successful operation'.[41]

Plans for a new Westminster Bridge were finally published and a controversy immediately began between James Walker and Charles Barry. It centred mainly on the shape of the arches, Walker favouring circular arches as in the old bridge, Barry pointed arches designed to harmonise with the new Palace of Westminster, then just beginning to rise from the ashes of the fire of 1834. Barry had also produced a design for a new superstructure for the old bridge, 'merely for the purpose of showing the possibility of raising a sightly superstructure in harmony with the intended new Houses of Parliament upon the present piers'.

But the committee, with the Speaker in the chair, had rejected this on the grounds that the work in progress was for repairs only. Even Sir Charles Eastlake entered the fray with a stately letter from the Royal Fine Arts Commission, 'on the importance of considering the effect which may be produced on the view of the new Houses of Parliament by the alterations about to be made in the structure of the bridge'. This was no affair of the Royal Fine Arts Commission, he hastened to say, but nevertheless, 'they venture to entertain a hope that the Trustees will give it their attentive consideration'.[42] The difference between Walker and Barry began with a slight divergence of opinion on a professional point, but rapidly developed into a bitter quarrel during which Barry wrote to the commissioners accusing Walker of 'endeavouring to defame my character in influential quarters by imputing improper motives to me . . . relative to the improvement of the superstructure'.[43] Walker and Burgess drew up a long memorandum which they presented to the Speaker, justifying their design and claiming that the pointed arch applied to bridges in mid-nineteenth century was an anachronism. Barry followed this with a letter to The Times quoting the best mathematical and engineering opinion of the day and no doubt lobbying for the support of Eastlake and the Royal Fine Arts Commission.

Their designs for the new bridge, together with others by George Rennie and Thomas Page (all dated between 1843 and 1846), were published in a Parliamentary Report of 1846, and it is of interest to note that Barry's published design, so far from being Gothic in style, consisted of 'five elliptical arches of a light and graceful structure'.[44]

In spite of all the effort the new designs were shelved. The committee advised the House of Commons that 'no case has been made out to justify the Committee in recommending to the House the pulling down the present Bridge and the constructing a new one'; though, curiously enough, nearly all the witnesses had stated their belief to the contrary— that the old bridge was too far gone and any further tinkering would be a waste of time.[45] Provided that money was made available, expert opinion was unanimously in support of a new bridge. As *Punch* declared the following year, 'this unhappy old structure which has long ago been condemned by its piers, has been the subject of a conversation in the Commons. We never see a loaded omnibus going over it without wondering whether it will get safe to the other side. . . The crazy old concern is past mending and the only remedy is that proposed in the House of Commons the other night, namely to pull it down and build a new one'.[46] Lord Lonsdale, a former Commissioner of Woods and Forests and a keen member of the select committee, tried to bring matters to a head. He voiced the general impatience and discontent by threatening to bring in a motion of no confidence in the engineer: 'that the Commissioners having been encouraged by the confident opinions given to them from time to time by Mr Walker to sanction the outlay of large sums of money in the expectation that the present Bridge could be repaired and made permanently secure, and as it has turned out that this is incorrect and injudicious professional advice, they now no longer adopt the suggestions and opinions of Mr Walker'. This motion was recorded in the journal but it was not raised at the next meeting, evidently being thought to be unjustifiably harsh, and Walker and Burgess continued to issue their weekly reports, often with special emphasis on the need for a new bridge and for a temporary bridge while the new one was being built.[47]

In 1846 another select committee (chairman Sir Harry Inglis) produced a combined report on Westminster Bridge and the New Palace of Westminster, having closely examined twelve witnesses in March and April and taken into consideration the earlier report of 1844.[48] The new report was a bulky document of over 200 pages and an extra forty pages of index. The situation was summed up in evidence by James Walker.

After the Bridge has sunk and twisted about in the way it has done, from the commencement of its building to the present time, I have seen enough of it not to risk anything like a professional opinion upon it [its stability]... it is like a patient whose constitution I did not make, which has been in the hands of the doctors from the day it was built to the present time.[49]

The select committee concluded unanimously that most of the witnesses agreed the foundations had been 'originally vicious', that no amount of repair would ever make it sound again, and that a new bridge should be built immediately. *The Times* reprinted the report a week after its publication, together with a leader expressing astonishment at all the fuss and hesitation. Repairs were estimated at £70,000. The highest estimate for a new bridge was £360,000, and £100,000 could be furnished from the Bridge Estates. A sum in simple arithmetic showed that an outlay by the nation of £190,000 was all that had to be sanctioned 'to complete a new and handsome bridge, to obviate a large annual loss, and to ensure safety to many thousands of lives'.[50]

In July 1846 *Punch* published a drawing called 'The Perilous Pass of Westminster Bridge', showing dare-devil traffic crossing a jagged mountain range. This was followed a few weeks later by 'The Lament of Westminster Bridge', six doggerel verses of which this is a fair sample:

> Hard is the fate of an infirm old pile
>   While daily sinking on a cold damp bed;
> If they don't move me in a little while.
>   I certainly shall tumble down instead.

Only one arch was navigable and *Punch* shows a drawing of a fleet of early Victorian paddle-steamers struggling to squeeze through.[51]

During the autumn of 1846 an enormous quantity of masonry was taken off the superstructure, and the whole of the balustrade and alcoves was replaced with a wooden fence. Timber centres were fixed under three of the arches and the surface of the road was lowered so far that it lay in a ten-foot-wide trough, four steps down from the pavement. The bridge had become what *The Observer* described as a crazy and tottering structure, 'with the mournful appearance of a total wreck unable either to float or sink'.[52] The three timber centres and ramshackle wooden fence can be seen in Henry Pether's painting of the bridge, ruinous and disconsolate in the moonlight and shamed by the fine new Palace of Westminster rapidly rising nearby (Plate 48). As a correspondent in *The Builder* said, the state of affairs was monstrous, 'with three arches blocked up in the middle of the Thames, a disgrace to the country—the centre arch of a metropolitan bridge kept up by a wooden leg'.[53] When the timber centres were removed in 1858 they were found to be so rotten that had the bridge really been settling they would no more have withstood the pressure 'than so many trusses of straw'.[54]

Though the bridge still carried over 63,000 people a day, its life was very nearly over. A poetaster wrote a swan song on the old bridge's behalf.

> Oh here I am as you may see
>   In misery I am quaking.
> They have bruised my back and broke my bones,
>   Good lawk! how I am shaking . . .
> I have beheld some glorious sights
>   Upon the noble river;
> Steam boats cutting up and down,
>   Boats and barges too, amazing.
> Rowing matches, Lord Mayor's shows,
>   And the Parliament Houses blazing.[55]

Then in 1852 *The Builder,* under the heading 'Alas! Westminster Bridge', complained of endless delay, unnecessary expense and nothing accomplished for years. If a severe winter should happen, with a heavy accumulation of ice, the timber shoring might easily be swept away and the whole bridge collapse into the river, almost inevitably leading to the fall of Blackfriars Bridge too. The article censured both the City authorities and the Bridge Commissioners for blunders and wastefulness which might have been allowed a century ago but were inexcusable in 1851—the year of the Great Exhibition, which boasted to the world the country's prowess in industrial skills. James Walker was also criticised for agreeing to repair the bridge in the first place in face of all the evidence. 'It is all very convenient to condemn Labelye', it continued, 'but in what respect is that engineer his superior who attempts that which, judging from the evidence of the first men of the day, could never succeed.'[56]

Finally a report of the commissioners appeared in 1852, declaring that a new iron bridge should be built at once, the old one being used as a temporary bridge in the meanwhile.[57] The committee added a rider that the dilapidated state of the old bridge, the uncertain expense of maintaining it even as a temporary structure and its inconvenience to both land and water traffic made it imperative that no time should be lost. In fact, they said, if work was to begin early in 1853, it could be open in less than two years. The report appeared on 1 April. A letter in *The Times* next day decided that it must be a hoax and recommended that the new bridge, should it ever materialise, ought to be called Pons Asinorum in honour of the day.[58]

The public had long since ceased to expect any unseemly haste in the affairs of Westminster Bridge and its last few years ran true to form. The corporation of the commissioners was dissolved in 1853 and the bridge estates were transferred to the Office of Woods and Public Buildings.[59] Tenders for the new iron bridge, designed in the Gothic taste by Thomas Page, were ordered to be delivered by 1 February 1854 and were to include the removal of the old bridge. The first elm pile was driven on the Westminster shore in July 1854. Shortly afterwards the contractors, Messrs Mare & Co., failed and the work was delayed for another two years, until Page took it on himself. In 1857 it was hoped to open half the new bridge, but in a long progress article a year later *The Times* regretted that these hopes were quite unfounded and 'there is every possibility of our being doomed to another twelve months of the ricketty dangerous old bridge which is every day becoming more obstructive, more unsightly, and if possible more ruinous'. Besides, the stench of the river was so appalling that the workmen at water level were frequently overcome with nausea and vomiting, and sometimes as many as six or seven a day had to be sent home. The sickness fund of twelve shillings a week was completely exhausted.[60]

Eventually half the new bridge was opened to traffic on 1 March 1860 the old bridge still being used for people on foot, on horseback or pushing barrows. The queen decided not to open it, and the ceremony was performed by Page and W. F. Cowper, Commissioner of Works, walking sedately over it, followed by a crowd of boys racing across at full speed. The crown of the old bridge could be seen ten feet above the parapet of the new one.[61]

Two years later, on Queen Victoria's birthday in 1862—in fact at 3.45 in the morning of 24 May, the precise moment of Her Majesty's birth— the new Westminster Bridge was opened to a salute of twenty-five guns in honour of the years of her reign. She had agreed to perform the ceremony in person but of course, as *The Times* remarked piously, 'since the lamented death of the Prince Consort (14 December 1861) this has been out of the question'. The opening attracted very little attention. People had long been accustomed to using one half only and in fact Page had been congratulated on being able to keep the traffic flowing throughout. It was quite a feat. 'When we see', said *The Times*, 'how week after week great main avenues of traffic are blocked up if a gas pipe happens to be out of order or a 3ft sewer·wants a few bricks in its crown, it is a contrast creditable to Mr Page that he has been able to build his bridge across the Thames without a single stoppage of traffic at any time.'[62]

The remains of 'the old invalid' were quickly hurried away. The great submerged Portland stone blocks of the piers were removed by cranes and by men working in the diving suits which had just superseded the old diving bell. They could stay under water for three hours in almost total darkness, fixing the cables to the lewises by sense of touch until their hands became so numb that they could strike them with hammers without noticing—until they came to the surface.

By the summer Labelye's bridge had disappeared and Page's new iron bridge stood alone. Apart from a few blocks of Portland stone in Page's abutments, nothing remained of old Westminster Bridge—Fielding's 'Bridge of Fools' and Labelye's 'very great Ornament to the Capital of the British Empire'—except a majestic memory immortalised in a sonnet by Wordsworth and the paintings of Canaletto and Samuel Scott.

## NOTES TO CHAPTER 18

1. *London Magazine* (1750) p 600.
2. *Gephyralogia, or a History of Bridges, Ancient and Modern* (1751)
3. Maitland *History of London* (1756) II p 1349 and repeated word for word in Dodsley *London and its Environs* (1761), *England Displayed* (1769), and several others.
4. Strype *Survey of London* (1754–5) II p 641.
5. Labelye *Description* pp 14 and 25–6.
6. Smollett *Travels through Fance and Italy* letter XXIX 20 February 1765.
7. *World* 27 September 1753. Tilt boats were large rowing boats covered with awnings (tilts) and plying between London and Gravesend.
8. J. T. Smith *Nollekens and his Times* 1829 (ed. Wilfred Whitten I pp 153–4). A play was said to have been performed, 'The Mystery of the Murder on Westminster Bridge'.
9. Pierre-Jean Grosley *A Tour to London or Observations on England and its Inhabitants* (1765).
10. C. P. Moritz *Journeys of a German in England in 1782.*
11. Stephen Riou *Short Principles for the Architecture of Stone Bridges* (1760) pp 90–1.
12. Grosley op cit p 27.
13. César de Saussure *A Foreign View of England* pp 196–8.
14. Letter from John Frost, secretary to the RHS , 19 May 1827.
15. *Low Life* (1764) p 72.
16. See Appendix page 290.

17. Bridge Minutes 20 March 1794, 26 April 1795 and 12 April 1797; Act 34 George III, c 51 (1794).
18. Bridge Minutes 10 June 1805.
19. Ibid 15 June 1824.
20. Ibid 7 April 1752.
21. Ibid March, April, May and 20 November 1753.
22. Ibid 2 April 1754.
23. *London Chronicle* 30 July 1759.
24. Bridge Minutes 4 and 11 December 1759.
25. *Gentleman's Magazine* (1760) pp 364–5.
26. Ibid (1761) p 610.
27. *London Chronicle* 17–19 April 1766, *Gazetteer and New Daily Advertiser* 19 April 1766, *Public Advertiser* 19 April 1766.
28. Bridge Minutes 26 April 1796. For a description of the inside of the hollow pier see *Gentleman's Magazine* June 1802 p 575.
29. Thomas Gayfere's manuscript Account Book belonged to Rennie and was given to the ICE by his widow after his death in 1821.
30. Bridge Minutes 11 June 1811, 15 May and 24 June 1812.
31. Ibid 30 May 1827.
32. Telford's report to Clementson in Bridge Minutes 12 May 1823 was printed in Appendix to Select Committee Report (1844) p 777.
33. Bridge Minutes 24 August 1831 and 15 July 1832.
34. Ibid 2 September 1835 and 25 June 1836.
35. Lord Lowther, Sir Henry Parnell, Sir Edward Knatchbull, John Ireland Blackburne and Joseph Hume. The Speaker joined the committee in 1840.
36. Bridge Minutes 20 July 1836.
37. James Walker's *Report on Westminster Bridge* 28 February 1837.
38. *The Times* 14 August 1846 5a.
39. 'These counter-arches evidently rest on false abutments pressing unequally up and distressing the arches which support them. . . .' The whole weight of the road pressed on one spot and Walker and Burgess relieved the strain by substituting a brick wall to carry the road over the vault (Committee Minutes pp 58 and 76).
40. Committee Minutes 25 May 1841 pp 72–7.
41. *The Times* 9 June 1843 7a.
42. Committee Minutes 19 July 1843.
43 Bridge Minutes 18 May 1843.
44. PP 1846 (477) VI. *The Times* 28 March 1844 8c.
45. Witnesses in 1844 were Clementson, Swinburne, Walker, I. K. Brunel, J. M. Rendel, George Rennie, William Cubitt (engineer), William Cubitt (contractor), Thomas Page, William Hosking, Charles Barry, Charles Nixon (superintendant under Walker and Burgess), James McAdam and Philip Hardwick (surveyor succeeding Craig).
46. *Punch* (1845) IX p 82.
47. Bridge Minutes 28 July 1845.
48. On this occasion the bridge witnesses were Walker, Rennie, Page and Pennethorne.

49. James Walker 19 May 1846 in *Report* para 1305.

50. *The Times* 14 August 1846.

51. *Punch* (1846) XI pp 56, 98, 114 and 154.

52. *The Observer* 18 October 1846.

53. *The Builder* 25 January 1851.

54. *The Times* 3 September 1858.

55. *The Downfall of Westminster Bridge* (c 1850) in a volume of Poetical Broadsides etc (BM 11621.K.5.370).

56. *The Builder* 16 August 1851.

57. An Act had been passed in 1850 authorising a temporary bridge to be built but this was shouted down as being an unwarranted expense (Private Act 13 Victoria 112).

58. *The Times* 2 April 1852 8f.

59. Acts 16 & 17 Victoria, c 46 (1853).

60. *The Times* 3 September 1858 7ef. *Punch* 3 July 1858 suggested that a suitable fresco for the new Houses of Parliament might be Father Thames introducing his offspring, diphtheria, scrofula and cholera, to the fair City of London.

61. *The Times* 5 March 1860 7c.

62. Ibid 27 February 1862 5e.

# Appendix

# KEY TO CHRONOLOGY

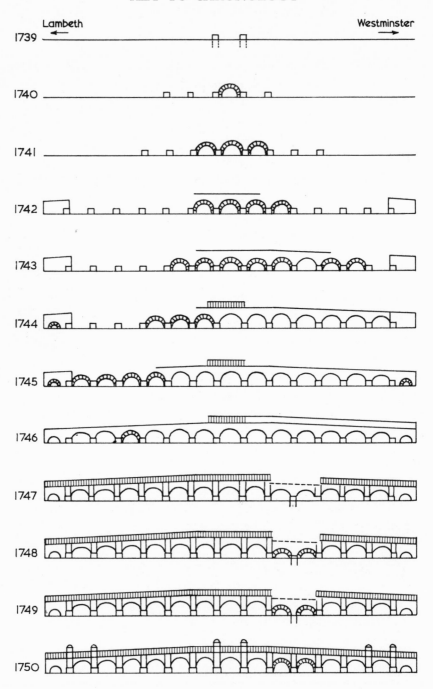

## MEASUREMENTS OF WESTMINSTER BRIDGE

| | | |
|---|---|---|
| Westminster abutment | 76 feet | |
| Westminster abutment arch | 25 | |
| Westminster abutment pier | 12 | |
| arch | 52 | |
| pier | 12 | |
| arch | 56 | |
| pier | 13 | |
| arch | 60 | |
| pier | 14 | |
| arch | 64 | |
| pier | 15 | (sinking pier) |
| arch | 68 | |
| pier | 16 | |
| arch | 72 | |
| west middle pier | 17 | |
| middle arch | 76 | |
| east middle pier | 17 | |
| arch | 72 | |
| pier | 16 | |
| arch | 68 | |
| pier | 15 | |
| arch | 64 | |
| pier | 14 | |
| arch | 60 | |
| pier | 13 | |
| arch | 56 | |
| pier | 12 | |
| arch | 52 | |
| Surrey abutment pier | 12 | |
| Surrey abutment arch | 25 | |
| Surrey abutment | 76 | |
| | ——— | |
| whole breadth of Thames | 1220 | |

| | | | |
|---|---|---|---|
| length of two abutments | 152 | | |
| section of fourteen piers | 198 | solids | 350 feet |
| span of fifteen arches | 870 | voids | 870 |
| | ——— | | ——— |
| | 1220 | | 1220 |

Roadway 30 feet, footways 6 feet each, parapets 1 foot each
Total width of bridge 44 feet.

# DATES OF PIERS, CENTRES AND ARCHES

## PIERS

| *Contract* | *Complete** |
|---|---|
| Two middle piers | 1 August 1739 |
| 22 June 1738 | (27 January 1742) |
| | |
| Abutments and two half-piers | |
| 23 May 1739 | |
| | |
| Four more piers | 3 June 1741 |
| 19 June 1739 | |
| | |
| Alterations to piers and abutments | |
| 3 July 1739 | |
| | |
| Six remaining piers | 27 October 1742 |
| 28 January 1741 | |

## CENTRES

| *Westminster* | *Contract* | *Built** | *Struck** |
|---|---|---|---|
| 25 | 11 January 1744 | | 19 January 1745 |
| 52 | 2 April 1742 | 6 December 1743 | 19 January 1745 |
| 56 | 2 April 1742 | 26 July 1743 | 26 February 1745 |
| 60 | 2 April 1742 | 28 May 1743 | 6 June 1744 |
| 64 | 2 April 1742 | 22 December 1742 (June 1748) | 23 May 1744 (June 1751) |
| 68 | 5 June 1741 | 30 June 1742 (June 1748) | 14 December 1743 (June 1751) |
| 72 | 30 June 1740 | 3 September 1741 | before February 1745 |
| 76 | 30 June 1740 | 11 June 1741 | 3 December 1744 |
| 72 | 30 June 1740 | 24 November 1741 | 15 October 1745 |
| | *ordered to be set up* | | |
| 68 | 1 February 1744 | | December 1745 |
| 64 | 23 May 1744 | | 7 September 1746 |
| 60 | 22 January 1745 | | 4 April 1747† |
| 56 | 22 January 1745 | | 19 December 1746 |
| 52 | 22 January 1745 | | 5 November 1746 |
| 25 | 11 January 1744 | | |
| *Lambeth* | | | |

*Labelye's or Graham's certificate
†'The last centre'—Bridge Minutes 7 April 1747

# ARCHES

| *Westminster* | *Contract* | *Complete\** |
|---|---|---|
| abutment | 23 May 1739 | February 1742(?) |
| 25 | 23 May 1739 | March 1744 |
| 52 | 2 April 1742 | 20 March 1744 |
| 56 | 2 April 1742 | 20 March 1744 |
| 60 | 2 April 1742 | 20 March 1744 |
| 64 | 2 April 1742 | 20 March 1744 |
| | 1741 | taken down May 1749 rebuilt 10 November 1750 |
| 68 | 1741 | 20 March 1744 |
| | 2 April 1740 | 15 January 1745 taken down May 1749 rebuilt 10 November 1750 |
| 72 | 2 April 1740 | 15 January 1745 |
| 76 | 2 April 1740 | 15 January 1745 turned February 1742 cornice 17 October 1744 balustrade September 1745 |
| 72 | 2 April 1740 | 15 January 1745 |
| 68 | 2 April 1740 | 15 January 1745 |
| | 31 August 1743 | 20 April 1745 |
| 64 | 31 August 1743 | 20 April 1745 |
| | 30 May 1744 | 17 March 1746 |
| 60 | 30 May 1744 | 17 March 1746 |
| 56 | 30 May 1744 | 18 November 1746 |
| | 14 May 1745 | |
| 52 | 14 May 1745 | 18 November 1746 |
| 25 | 23 May 1739 | |
| abutment | 23 May 1739 | February 1742 |
| *Lambeth* | | |

\*Date of Labelye's or Graham's certificate

# CONTRACTS

BRIDGE CONTRACTS I, JUNE 1737-FEBRUARY 1744 (PRO WORKS 6/42)

(1)  22 June 1737, p 1. Agreement with the Bank of England (James Collier and John Waite, cashiers) for receiving Lottery money.

(2)  19 May 1738, p 4. Commissioners' resolution to employ Jelfe and Tufnell.

(3)  9 June 1738, p 5. Bond between Jelfe and Tufnell and their securities John Lawton and Lancelot Storey, and Sundon and Wager.

(4)   22 June 1738, p 8. Contract between Ayloffe and Blackerby and Jelfe and Tufnell for the two middle piers (Labelye's completion certificate 27 January 1741/2).

(5)   9 June 1738, p 16. Bond between King and Barnard and their securities William Barnard, tallow merchant of St Katherine's near the Tower, George Fryer, butcher of Kensington, Joshua Ransom and John Smith, merchants of London (£4,000), and Sundon and Wager.

(6)   22 June 1738, p 19. Contract between King and Barnard and Ayloffe and Blackerby for building the wooden superstructure (contract cancelled 9 July 1740).

(7)   12 July 1738, p 19. Lease from Ransom and Smith, timber merchants of Lambeth, to Richard Graham for part of their timber-yard at Stangate.

(8)   11 September 1738, p 31. Contract with Robert Halliwell of St Martin in the Fields for three horses to work the pile-driving engine.

(9)   6 October 1738, p 34. Contract with Hassall Short for his house and yard next to Ransom & Smith's yard, together with free use of crane, sawpit, wharf, sheds, etc, occupied by Richard Crooke.

(10)   12 October 1738, p 38. Contract with Robert Smith, ballast man of Old Brentford, to level the foundation of the first pier, employing two gangs of men, in forty-two days.

(11)   17 October 1738, p 40. Agreement with Hercules Taylor, carpenter of St Margaret's, to lease at 4s 6d a week his land and dwelling house at the Woolstaple.

(12)   13 February 1739, p 43. Agreement with Robert Smith to level ground for the second pier in six weeks.

(13)   15 May 1739, p 45. Contract with Robert Halliwell for three horses to drive piles for three months.

(14)   23 May 1739, p 47. Bond between Jelfe and Tufnell and Ayloffe and Blackerby for building abutments and two half piers.

(15)   23 May 1739, p 51. Contract for the above.

(16)   19 June 1739, p 63. Bond and articles for building four more piers (Graham's certificate of completion in Bridge Minutes 3 June 1741, Labelye's 17 January 1742).

(17)   3 July 1739, p 77. Agreement with Jelfe and Tufnell for alterations to abutments and piers as directed by Labelye.

(18)   1 June 1739, p 78. Agreement with Robert Smith to level ground for the third pier in six weeks.

(19)   15 August 1739, p 80. Contract with Halliwell for three horses etc.

(20)   27 August 1739, p 82. Agreement with Smith for fourth pier.

(21)   18 October 1739, p 84. Agreement with Smith for fifth pier.

(22)   10 March 1740, p 86. Contract with Halliwell for three horses etc.

(23)   2 April 1740, p 88. Bond between Jelfe and Tufnell and their securities John Lawton and Edmund Holmes, soap-maker of St Margaret's, for the three middle arches.

(24)   2 April 1740, p 93. Contract for the three middle arches and part of the next arches (Labelye's certificate 7 January 1744/5).

(25)  27 May 1740, p 102. Agreement with Smith for sixth pier.

(26)  10 June 1740, p 104. Contract with Halliwell for three horses etc.

(27)  30 June 1740, p 106. Bond from James King and the Earl of Pembroke (£1,000) to Ayloffe and Blackerby for centres for turning the three middle arches.

(28)  30 June 1740, p 110. Articles of agreement for the three centres.

(29)  21 July 1740, p 119. Agreement with Robert Smith to level ground for the Surrey abutment.

(30)  13 August 1740, p 121. Bill of sale from King and Barnard to Richard Graham of timber provided for the naked superstructure.

(31)  13 August 1740, p 127. Release from King and Barnard to Ayloffe and Blackerby.

(32)  10 September 1740, p 129. Contract with Halliwell for three horses etc.

(33)  10 October 1740, p 131. Agreement with Smith for seventh pier.

(34)  27 December 1740, p 133. Contract with Halliwell for three horses etc.

(35)  28 January 1741, p 135. Bond from Jelfe and Tufnell for the six remaining piers (£5,000).

(36)  28 January 1741, p 139. Agreement for the six remaining piers (Graham's certificate 27 October 1742, Labelye's 24 November 1742).

(37)  9 February 1741, p 148. Agreement with Smith for the eighth pier.

(38)  27 March 1741, p 150. Contract with Halliwell for three horses etc.

(39)  11 May 1741, p 152. Agreement with Smith for the ninth pier.

(39)  11 May 1741, p 152. Agreement with Smith for the ninth pier.

(40)  5 June 1741, p 154. Bond from King and John Jones, ironmonger of St Martin in the Fields (£400), to build the fourth centre.

(41)  5 June 1741, p 158. Agreement for the fourth centre (Graham's certificate 16 June 1742).

(42)  27 June 1741, p 169. Contract with Halliwell for three horses etc.

(43)  4 August 1741, p 171. Agreement with Smith for the tenth pier.

(44)  28 September 1741, p 173. Contract with Halliwell for three horses.

(45)  16 November 1741, p 175. Agreement with Smith for the eleventh pier.

(45a) 18 November 1741, Bridge Minutes. Contracts with Henry Conyers for vaulting under Bridge Street and with John Wilkins for paving Bridge Street and footway.

(46)  15 February 1742, p 177. Halliwell for pile-driving horses.

(47)  8 March 1742, p 179. Agreement with Smith for the twelfth pier.

(48)  2 April 1742, p 181. Bond from King and Joshua Ransom (£400) to build the four centres next to the Westminster abutment.

(49)  2 April 1742, p 184. Agreement for the four centres.

(50)  2 April 1742, p 194. Bond from Jelfe and Tufnell for the four arches and part of an arch and piers next to the Westminster shore... two years.

(51)  2 April 1742, p 198. Agreement for the above (Labelye's certificate 20 March 1744).

(52)  15 May 1742, p 207. Halliwell for pile-driving horses.

(53)  30 June 1742, p 209. Agreement with Jelfe and Tufnell for torus, cornice, towers and footway over the middle arch (Labelye's certificate 18 October 1744).

(54)   16 August 1742, p 216. Halliwell for pile-driving horses.

(55)   23 August 1742, p 218. Agreement with Smith for the thirteenth pier.

(56)   25 August 1742, p 220. Contract with William Bowen for fourteen circular wedges for discharging the centres.

(57)   16 February 1743, p 227. Halliwell for pile-driving horses.

(58)   17 May 1743, p 229. Halliwell for pile-driving horses.

(59)   1 June 1743, p 231. Agreement with Smith for the fourteenth and last pier.

(60)   18 August 1743, p 233. Halliwell for pile-driving horses.

(61)   31 August 1743, p 235. Contract with Jelfe and Tufnell for the east 68ft and part of east 64ft arches.

(62)   31 August 1743, p 244. Bond for the above.

(63)   19 November 1743, p 247. Halliwell for pile-driving horses.

(64)   23 November 1743, p 249. Contract with Jelfe and Tufnell for the cornice and parapet walls over the octagon towers and for paving the recesses (Labelye's certificate 22 April 1746).

(65)   23 November 1743, p 256. Bond for the above.

(66)   27 February 1744, p 259. Contract with Charles Marquand for cutting off piles under the centres (contract cancelled 18 April 1744).

(67)   27 February 1744, p 265. Bond for the above.

(68)   20 February 1744, p 268. Halliwell for pile-driving horses.

BRIDGE CONTRACTS II, MARCH 1744—MARCH 1750 (PRO WORKS 6/48)

(69)   21 March 1744, p 3. Contract with Jelfe and Tufnell for plinths, balustrades, pedestals and rail over the middle arch . . . four months (Labelye's certificate 6 August 1745).

(70)   21 March 1744, p 10. Bond for the above.

(71)   21 May 1744, p 13. Halliwell for pile-driving horses.

(72)   30 May 1744, p 15. Contract with Jelfe and Tufnell for the east 64ft and 60ft arches and part of an arch . . . six months after centre is ready (Labelye's certificate 17 March 1746).

(73)   30 May 1744, p 27. Bond for the above (£4,000) Barwell Smith and Francis Tregagle.

(74)   22 August 1744, p 31. Halliwell for horses.

(75)   21 February 1745, p 33. Halliwell for horses.

(76)   14 May 1745, p 35. Contract with Jelfe and Tufnell for the remainder of the east 56ft and 52ft arches . . . six months.

(77)   14 May 1745, p 47. Bond for the above (£4,000) Smith and Tregagle.

(78)   22 May 1745, p 51. Halliwell for horses.

(79)   17 December 1746, p 53. Contract with Jelfe and Tufnell for the remainder of the torus, cornice, plinths, baluster, pedestals, rail and footways . . . over west 56ft and 52ft arches . . . three months.

(80)   17 December 1745, p 66. Bond for the above (£500) Smith and Tregagle.

(81)   5 August 1746, p 70. Robert Smith for furnishing Thames ballast for the new road in Surrey.

(82)   5 August 1746, p 75. Bond for the above.

(83)  18 September 1746, p 78. Contract with Smith for furnishing sand and removing earth from the new road.

(84)  18 September 1746, p 87. Bond for the above.

(85)  21 February 1747, p 88. Contract with Jelfe and Tufnell for twelve semi-octagonal domes over the recesses . . . twelve months.

(86)  21 February 1747, p 99. Bond for the above.

(86a) 2 March 1749, Bridge Minutes 14 March. Agreement with John Smith of Lambeth to let two cranes and a wharf near Stangate.

(86b) 6 March 1749, Bridge Minutes 14 March. Agreement with Samuel Price to carry stone from the bridge to the wharfs.

(87)  1 July 1749, p 102. Halliwell for pile-driving horses.

(88)  6 February 1750, p 106. Contract with Jelfe and Tufnell for taking down and rebuilding the damaged pier and arches.

(89)  2 March 1750, p 120. Agreement with John Smith to let his crane and wharf at Stangate for another year.

(89a) 19 April 1751, Bridge Minutes 23 April. Agreement with Joseph Liddall, carpenter of St George's, for sale of the centres under the west 68ft and 64ft arches and counter-arch, scaffold, piles, booms etc.

## BRIDGE BUILDERS

### OFFICERS

Sir Joseph Ayloffe, principal clerk to the commissioners
John Bowack and John Lewis, assistant clerks
Charles Labelye, engineer
Nathaniel Blackerby, treasurer 1736–42
Samuel Seddon, treasurer 1742–50
Richard Graham, surveyor and comptroller of works 1738–49
Obadiah Wylde, surveyor and comptroller of works 1749–50
Thomas Lediard (senior), surveyor and agent 1736–43
Thomas Lediard (junior), surveyor and agent 1743–50

### TRADESMEN AND PROFESSIONALS

| | |
|---|---|
| Attorneys | Taylor White, George Monkman, Mark Frecker, John Nethersold, Hugh Watson. |
| Bailiffs | John Lever (high bailiff), Samuel Baldwin (deputy bailiff). |
| Ballast men | Robert Smith, Joseph Napper, Thomas Wootton, William Godfrey, William Westcombe, John Higgs. |
| Barge-masters | Samuel and Mary Price. |
| Bargemen and watermen | George Field, — Cornwall, — Babb, Thomas Greenway, Samuel Horlock. |
| Barges | *Rainbow, Robert, Richard, New Bridge.* |
| Blacksmiths | Benjamin Holmes, Joseph Pattison, Henry James, Thomas Wagg. |
| Bookbinders | Thomas Corble, Thomas Delafield, John Walter, Joseph Smith. |

| | |
|---|---|
| Bricklayers | Henry Conyers, — Drake, Richard Fortnum, Stenton, Vail, John Chidley. |
| Brickmaker | Thomas Scott. |
| Cabinet-maker | William Gomm (Bridge Office furniture). |
| Carpenters | King & Barnard, William Etheridge, Chandler, Pope, Ruby, Stedman, James Neeld, John Simpson. |
| Carters | Robert Halliwell, John Hatchings, John Mears, Edward Sexton, — Huddle, Thomas Cosby, Stephen Burgess. |
| Cleaner | Obadiah Richards. |
| Clerks | Zachary Chambers, Zachary Hamblyn, Samuel Lewis, John Lewis, William Gordon, Robert Raser. |
| Coal merchants | Samuel Price, John Wright. |
| Cryer at Court | William Gordon. |
| Dredgers | Edward Nicholson, Robert Smith, Joseph Napper, John Simpson (see labourers). |
| Doorkeepers and messengers | John Grandpré, Robert Pealing, John Lewis. |
| Ferryman | Henry Folkes. |
| Fisherman | James Parsons. |
| Glaziers | John Graham, John Pilmer, Richard Mims, Robert Jesson. |
| Housekeepers | Mary Cundhill, Elizabeth Dagge, Martha Pye. |
| Innkeeper | John Stephenson. |
| Ironfounders | William Bowen, Dominick Donnelly, Joseph Hamilton. |
| Jury Attendants | William Goosetrey, William Patrick (deputies to the high baliff). |
| Labourers digging the foundations | Isaac Barnett, Thomas Barley, Frank Bates, John Bellamy, Thomas Burley, Isaac Cornwall, Edward Davis, John Dennison, Thomas Eames, Isaac Gardner, Thomas Gordon, Samuel Hallett, Robert Hammond, William Hammond, Joseph Hatfield, Edward Hazey, Daniel Hind, William Holmes, James Isaacs, Robert Jackall, Richard Joy, Matthew Kelly, Pat Kelly, Thomas Miles, Abraham Mouldy, Joseph Napper (foreman), John Newton, William Nichols, John Riddell, Richard Ridgel. Henry Rose, Jonathan Sykes, John Smith, Sam. Taylor, Rufus Tufton, Sam Twille, Sam Walters, Pat Walters, James Wenton, James Wigball, Edward Winter. |
| Lamplighters | Thomas Walker, Mrs Walker, Jonathan Durden, Balchin Walker & Search, Richard Walker, David Ross, Elizabeth Ross, Joseph Hayling, Featherstone, Gas Light Company. |
| Lampmakers | Daniel Cuthbertson, Thomas Walker, Mrs Walker. |
| Lighterman | Hugh Griffith. |
| Lottery adviser | Richard Shergold. |

| | |
|---|---|
| Masons | Jelfe & Tufnell, William Fellows, Thomas Gayfere. |
| Mathematical instrument maker. | Thomas Heath. |
| Ordnance storesman | George Campbell. |
| Painters | Thomas Etheridge, John Hughes, William Pickering, William Reynolds, Jonathan Brewster. |
| Paviours | John Wilkins, Thomas Phillips, Edward Mist. |
| Pile-drivers | Robert Halliwell, Charles Marquand, William Etheridge, Edward Ruby. |
| Plumbers | Edward Ives, Edward Jones, Thomas Shepherd, George Devall, Richard Troubridge. |
| Printers | John and Robert Baskett, Samuel Richardson. |
| Pumpmakers | John Broadbent, Sam and John Garey. |
| Quarrymen | Thomas Roper, Tucker bros., T. Bryer, Edward Tizzard, Captain Tarzell. |
| Roadmakers | Henry Monk, Thomas Phillips, John Simpson, William Godfrey. |
| Ropemaker | William Stubbs. |
| Sailmaker | William Crandly. |
| Sawyers | David Jenkins, Robert Pealing. |
| Sewermen | Edward Harris, John Mears |
| Shipwright and boatbuilder | Thomas Patten. |
| Scull and oar maker | Robert Metcalf. |
| Stationers | Thomas Gamull, Robert Powney. |
| Stonecarvers | Thomas Evans, George Murray, (Sir Henry Cheere). |
| Surveyors | James Stedman, John Wilkie, Thomas Fearson, Richard Graham, Thomas Hinton, — Horne, Thomas Lediard (father and son), Captain John Mitchell, Thomas Walker (surveyor of Crown lands). |
| Tallow chandlers | John Craister, John Sheffield. |
| Timber merchants | William Godfrey, William Pacey, Leonard Phillips, John Smith, John Spencer, Edward and Richard Steavens, Kemble Whatley, Edward Ruby. |
| Upholsterer | Joseph Nicolson (Bridge Office furniture). |
| Vestrymen | Sir Thomas Crosse, Dr Richard Lilly, Peter Buscarlet, James Neeld, John Machan. |
| Watchmaker | James Vauloué. |
| Watchmen (1750) | Westminster: Anthony Gardiner, Robert Thorn, John Lockyer, Thomas Woodyer, Robert Williamson, Henry Sims. Lambeth: William Page, Thomas East, Francis Smith, Charles Heather, John Oldfield, Thomas Cross. |
| Wharf owners | Joshua Ransom, John Smith. |
| Wheelbarrow makers | Matthew Pope, John Walker. |

### LATER OFFICERS

| | |
|---|---|
| Treasurer and | Samuel Seddon (1742–79), Thomas Ludbey (1779–91) |

| | |
|---|---|
| secretary | John Clementson (1791–1805), John Clementson, junior (1805–    ). |
| Surveyor and agent | John Simpson (1754–74), William Jewitt (1774–1801), Charles Craig (1802–29). |
| Clerk of the Works | William Swinburne (1823–    ). |

## ABBREVIATIONS USED IN THE NOTES

| | |
|---|---|
| Bridge Accounts | Accounts of Receipts and Payments 1737–50. |
| Bridge Committee of Accounts | Two volumes of proceedings of the Committee of Accounts, 1738–44 and 1745–51. |
| Bridge Contracts | Two volumes of contracts, 1737–50. |
| Bridge Minutes | Eleven volumes of proceedings of the Bridge Commissioners, 1736–1853. |
| CJ | House of Commons Journals. |
| Colvin *Dictionary* | H. M. Colvin *Biographical Dictionary of English Architects* 1660–1840 (1978). |
| Constable *Canaletto* | W. G. Constable *Canaletto* (1976 edition revised by J. G. Links). |
| Finberg *Canaletto* | Hilda Finberg 'Canaletto in England', *Walpole Society Journals* IX and X (1922–3). |
| GLC | Greater London Council, library and record office in County Hall. |
| ICE | Institution of Civil Engineers, London. |
| James *Short Review* | John James *A Short Review of the Several Pamphlets and Schemes . . . Westminster Bridge* (1736). |
| Labelye *Description* | Charles Labelye *A Description of Westminster Bridge . . . with an Appendix* (1751). |
| Labelye *Present State* | Charles Labelye *The Present State of Westminster Bridge . . . in a Letter to a Friend* (1743). |
| Labelye *Short Account* | Charles Labelye *A Short Account of the Methods . . . Foundations of the Piers of Westminster Bridge* (1739). |
| Lediard *Observations* | Thomas Lediard *Some Observations on the Scheme offered by Cotton and Lediard . . .* (1738). |
| PRO | Public Record Office, London. |
| RIBA | Royal Institute of British Architects, London. |
| *Short Narrative* | *A Short Narrative of the Proceedings of the Gentlemen concerned in obtaining the Act for building a Bridge at Westminster . . .* (1737). |
| Tradesmen's Accounts | Four volumes of Accounts, 1736–52. |
| *Vertue Notebooks* | George Vertue's Notebooks in *Walpole Society Journals* (1930–55). |
| VCH | The Victoria County Histories. |
| Wilton Papers | The Earl of Pembroke's papers at Wilton House. |
| Works Minutes | Two volumes of Works Committee Minutes, 1737–44 and 1744–51. |

# Glossary

(Technical words used in the commissioners' Minute Books, with modern spelling in brackets)

*Abutment*   The extremity of a bridge where it joins or abuts on the land.

*Arch*   An opening through which water flows—circular, elliptical, cycloidal, catenarian, etc.

*Archivolt*   The curve formed by the upper side of the voussoirs.

*Ballast*   Rough broken stone used as filling.

*Balustrade*   Parapet of balusters carrying a rail or coping.

*Banquet*   The raised footpath along the sides of a bridge.

*Battardeau (batardeau), coffer-dam*   Watertight enclosure of piles to enable piers to be built on a dry foundation.

*Bridging*   Transverse timbers used for strengthening.

*Caisson, cassoon*   Chest or flat-bottomed boat in which a pier is built.

*Campshed (camp-shed)*   Planks and sheet-piling used to contain earth at the landing-stages.

*Centre, centring*   Timber frames on which the arches were built. Throughout the commissioners' records the old spelling 'center' was used.

*Chatts*   Fine stone (?) 878 loads of screened gravel and chatts were laid on the roadway of the bridge in 1786.

*Chest*   see *caisson*.

*Coffer-dam*   see *battardeau*.

*Console*   Bracket.

*Counter-arch*   Concealed arch designed by Stukeley to relieve the weight on the sinking pier. The crown was too near the surface and it was removed with the accompanying quarter-arches in 1841.

*Cornice*   Projecting horizontal course along the base of the parapet.

*Crab*   A device for hoisting stone.

*Drift, shoot or thrust*   The force exerted by an arch along the length of a bridge.

*Elevation*   The projection of a bridge on the vertical plane, parallel to its length.

*Extrados*   The exterior curve of an arch.

*Foundations*   The bases on which piers are built. Labelye's foundations were wooden frames resting on the river-bed.

*Haunch*   see *reins*.

*Impost*   The surface of the pier from which an arch springs.

*Joggle*   A projecting tongue of stone used to join another stone.

*Keystone*   The middle voussoir.

*Kingpost*   A strong upright tie-beam used in the centre.

*Lewis*   A device used to hoist stone.

*Parapet*   The breast wall on a bridge to safeguard passengers.

*Pier*   Wall built for the support of an arch. Most piers were hollow but Labelye's piers were of solid Portland stone.

*Piles*   Timbers driven into the bed of a river, usually of elm, oak or fir, and shod with an iron point.

*Pile-driver*   An engine for hammering piles into the river bottom. The Westminster pile-driver was invented by a watch-maker, Vauloué, and driven by three horses.

*Pitch*   The vertical height of an arch from impost to keystone.

*Plan*   Projection on a plane parallel to the horizon.

*Plinth*   Projecting bases on which the octagonal towers rested.

*Puncheon* (1) dwarf post or stud used on the centres.

(2) timber cask sunk into the river-bed to support the feet of the centres.

*Racked*   Stretched and distorted by strain.

*Recess*   Alcove for pedestrians to shelter from traffic. Westminster Bridge had twenty-eight recesses, twelve of them covered.

*Reins*   The lower part of an arch. Labelye's secondary arches were thicker at the reins than at the crown.

*Relieving arch*   see *counter-arch*.

*Rib*   Timber arch of the centre. King's centres had five ribs.

*Rib wall*   Interior brick wall along the piers.

*Ring of an arch*   The voussoirs.

*Salient angle*  The projection at the end of a pier to divide the stream.

*Scheme arch* (*skeen arch*)  An irregular arch with segment less than a semi-circle.

*Shoot*   see *drift*.

*Sodding*   Soaking with water.

*Soffit, Sophet*   The underside of an arch.

*Spalts* (*spalls*)   Splinters of stone forced away by frost or weight.

*Spandrel*   Triangular space between arch and parapet, in Westminster Bridge built of Purbeck stone.

*Springer*   The lowest stones of an arch bearing on the impost.

*Sterling* (*starling*)   A case surrounding a pier for protection.

*Stilts*   A set of piles on which a pier is built. Labelye referred disparagingly to London Bridge being built on stilts.

*Tarris* (*tarras*)   A strong cement from Holland used under water on the piers.

*Tenterhooks*   Iron hooks along the temporary fencing to keep off intruders.

*Thrust*   see *drift*.

*Toom*   To smooth the stone with a chisel.

*Torous* (*torus*)   Convex moulding along the parapet.

*Voussoirs*   Wedge-shaped stones of an arch.

# Bibliography

## I. MANUSCRIPT SOURCES

BRITISH MUSEUM

Andrews Jelfe's Accounts and Letter Book (Add.MS.27587). Mostly letters to Roper about the delivery of stone from Portland, 1734–44.

GREATER LONDON COUNCIL

Fulham Bridge Minute Book, 11 January 1739 to 7 July 1770.
Westminster Bridge, draft articles of agreement, 1739–43.

GUILDHALL

Journals of the Common Council.
*Remembrancia* IX and MS.119.8. City opposition to Westminster Bridge. Map Case 141. City property in Westminster to be granted to the commissioners, W. Walker surveyor, 4 March 1741.

HOUSE OF LORDS, RECORD OFFICE

B112/1–44. Proceedings of the Commissioners for Westminster Bridge, 1738–91 (forty-four vols, copies of PRO Bridge Minutes).
B113/1–44. Treasurer's Accounts to the Commissioners for Westminster Bridge, 1738–91, and Rent Rolls 1764–80 (forty-four vols).
B114/1–12. Contracts made by the Commissioners for Westminster Bridge, 1738–1750 (twelve vols).

INSTITUTION OF CIVIL ENGINEERS

Thomas Gayfere's *Account Book* containing thirty-seven watercolour drawings probably intended to be used as illustrations in Labelye's book. It belonged to George Rennie and was presented to the ICE by his widow.
Labelye, *A Short Account*, containing Gayfere's drawings.
Francis Whishaw, typescript *Description of Westminster Bridge,* a shortened version published in *Proceedings of ICE*, I, 1838.

MIDDLESSEX COUNTY RECORD OFFICE

Middlesex Land Register.

OXFORD, BODLEIAN LIBRARY

Some particulars of Westminster Bridge, the Eddystone Lighthouse and Ramsgate Harbour.
William Stukeley's drawing for repairs to the sinking pier, 1748.

PARIS, ECOLE NATIONALE DES PONTS ET CHAUSSEES
'Collections des Ponts étrangers', sheet 9 Westminster Bridge, elevation, plan
  and section in colour.


PUBLIC RECORD OFFICE

| | | | |
|---|---|---|---|
| 6/28 | Commissioners' Minute Book, | | June 1736—February 1740 |
| 6/29 | ,, | ,, | February 1740—August 1741 |
| 6/30 | ,, | ,, | October 1741—January 1743 |
| 6/31 | ,, | ,, | January 1743—May 1745 |
| 6/32 | ,, | ,, | July 1745—December 1747 |
| 6/33 | ,, | ,, | January 1748—June 1751 |
| 6/34 | ,, | ,, | June 1751—February 1758 |
| 6/35 | ,, | ,, | February 1758—December 1767 |
| 6/36 | ,, | ,, | January 1768—March 1791 |
| 6/37 | ,, | ,, | April 1791—June 1829 |
| 6/38 | ,, | ,, | June 1829—July 1853 |
| 6/39 | Works Committee Minutes Book, | | August 1737—September 1744 |
| 6/40 | ,, | ,, | September 1744—April 1751 |
| 6/41 | Accounts Committee Minute, vol I, 1738–44 | | |
| 6/42 | Contracts, vol I, June 1737—February 1744 | | |
| 6/43 | Committee of Commissioners, Minute Book 1836–47 [sic] | | |
| 6/44 | Tradesmen's Accounts, 1736–41 | | |
| 6/45 | ,, | ,, | 1741–4 |
| 6/46 | ,, | ,, | 1744–9 |
| 6/47 | ,, | ,, | 1749–52 |
| 6/48 | Contracts, vol II, March 1744—March 1750 | | |
| 6/49 | Accounts Committee Minutes, vol II. 1744–51 | | |
| 6/50 | Sub Accounts, 1738–40 | | |
| 6/51 | ,, | 1740–3 | |
| 6/52 | ,, | 1743–9 | |
| 6/53 | ,, | 1750–7 | |
| 6/54 | Accounts of Receipts & Payments, 1737–50 | | |
| 6/55 | Warrants for Payments, 1737–43 | | |
| 6/56 | ,, | ,, | 1743–9 |
| 6/57 | ,, | ,, | 1749–59 |
| 6/58 | ,, | ,, | 1759–76 |
| 6/59 | Warrants and Receipts, 1740–1 | | |
| 6/60 | Register of Estates purchased by the Commissioners, 1739–47 | | |
| 6/61 | Abstract with plans of leases granted, 1749–98 (compiled c 1807) | | |
| 6/62 | Valuations, precepts, verdicts, etc 1739–46 | | |
| 30/26/12 | Fulham Bridge Minute Book I, | | 26 July 1726—11 November 1729 |
| 30/26/13 | ,, | ,, | II, November 1728—October 1736 |
| (G.L.C. | ,, | ,, | III, January 1739—July 1770) |
| 30/26/14 | ,, | ,, | IV, October 1770—1804 |
| 30/26/15 | ,, | ,, | V, 1805–25 |
| 30/26/16 | ,, | ,, | VI, 1826–61 |
| 30/26/17 | ,, | ,, | VII, 1861–82 |

WESTMINSTER CITY RECORDS
Rate Books, Poor Relief Ledgers
Burgess Court Records
Vestry Minute Books of St Margaret's and St John's

WILTON HOUSE
Letters from Ayloffe, Labelye, Lediard, Graham, etc to 9th Earl of Pembroke, 1746–50.


# II. PRINTED SOURCES

OFFICIAL JOURNALS, REPORTS, ACTS, ETC.
Journals of the House of Commons. Annual and Sessional volumes in the House of Commons library, abbreviated to CJ.
Journals of the House of Lords. Annual and Sessional volumes in the House of Lords library.

*Public General Acts*
Acts relating to Westminster Bridge and its approaches:

| 1736 | 9 George II, | c 29 | 1742 | 15 George II, | c 26 |
|---|---|---|---|---|---|
| 1737 | 10 | c 16 | 1744 | 17 | c 32 |
| 1738 | 11 | c 25 | 1745 | 18 | c 29 |
| 1739 | 12 | c 33 | 1756 | 29 | c 38 |
| 1740 | 13 | c 16 | 1757 | 30 | c 34 |
| 1741 | 14 | c 40 | (1814 | 54 George III, | cxxxii) |

*Later Acts*
1850  13 & 14 Victoria cxii, temporary bridge.
1852–3  16 & 17 Victoria, c 46, transfer of bridge estates to Commissioner of Works and Public Buildings.
1859  22 Victoria, c 19, additional space for Westminster approaches.
1864  27 & 28 Victoria, c 88, better regulation of traffic.

*Reports*
1759  Dance, Taylor, Phillips and Keene, MS report in Bridge Minutes.
1811  Rennie, MS report in Bridge Minutes.
1823  Telford, report printed in PP 1844 (477) VI 777.
1825  Hollingsworth, MS report in Bridge Minutes.
1827  Swinburne, MS report in Bridge Minutes.
1829  Telford and Clementson, PP 1844 (477) VI 780.
1831  Swinburne and Cubitt, ibid 782–4.
1835–6  Cubitt, ibid 785–7.
1837  Walker, plan to secure foundations, ibid 764.
1837  Cubitt, report on Walker's plan, ibid 769.
1839  Select committee's report.

1843  Walker and Burgess, Cubitt, ibid 789.

1844  Select committee's report containing various reports, list of commissioners, rent roll of estates, and a return of money spent on the bridge for repairs, alterations and professional services, 1810–38.

1845  Walker & Burgess, Mylne, Tierney Clarke, MS reports in Bridge Minutes.

1846  Select committee on the present state of Westminster Bridge and the Palace of Westminster with accounts of management and expenses, PP 1846 (177.349.574) XV.

1847  Walker & Burgess, MS report in Bridge Minutes.

1849  Swinburne, Burgoyne, MS report in Bridge Minutes.

1850  Select committee on Westminster Temporary Bridge Bill, PP 1850 (609) XIX 655.

1852  Letter from James Walker on the present condition of Westminster Bridge, PP 1852 (414) XLII 295.

1852  Commissioners appointed to consider the best and most convenient site for a new bridge, PP 1852 (1457) XVIII 605.

1852–3  Select committee, PP 1852–3 (622) XXXIX 467.

1856  Select committee, PP 1856 (389) XIV 225.

1857  Select committee, PP 1857 (246 sess. 2) IX 703.


*Accounts and Papers*

1854  Correspondence, tenders, contracts, PP 1854 (335) LXVII 407.

1854–5  Engineers' Report PP 1854–5 (347) LIII 407.

1856  „  „  PP 1856 (318) LII 205.

1861  Expenditure, PP 1861 (396) XXXV 531.

1862  „  PP 1862 (490) XXXI 585.


PAMPHLETS

*Abstract of the ACT with terms of Insurance for Adventurers in the Lottery,* 1736.

*Conjectures as to the most proper Place, between Scotland Yard and Vauxhall, for erecting a Bridge cross the River Thames from Westminster to the opposite Surrey Shore,* 1736.

Hales, Dr Stephen, *Proposals for making dams for Westminster Bridge,* 4 June 1737, MS addressed to Admiral Wager (WCL 942–132).

Hawksmoor, Nicholas, *A Short Historical Account of London Bridge with a proposition for a new stone-bridge at Westminster,* 1736.

James, John, *A Short Review of the Several Pamphlets and Schemes . . . a Bridge at Westminster,* 30 July 1736.

Labelye, Charles,

1735  Dissertation concerning the Paralogisms of F. W. Stubner relating to the Forces of Foreign Bodies in Motion. Letter to J. T. Desaguliers, 15 April 1735 (J.T.D., *Course of Experimental Philosophy,* 1745, II, p 77). Draughts and Descriptions of Richard Newsham's Fire Engines, J. T. Desaguliers, op cit., pp 505–19.

1736  *Map of a Proposed new Harbour at Sandwich,* 1736 (Kent Archives Office, Maidstone, Sa/P1).

1736  *A Mapp of the Lands situated between the Town and Port of Sandwich and the Sea Shore, and between the River Stower and the Road to Deal . . . from the Actual Survey taken in November and December 1736 by Charles Labelye* (Photostat in BM Maps 188.S.2 (6)).

1737  *A Mapp of the Downes . . . by Charles Labelye, Engineer, December 1736.* London 1737.

1739  *A Short Account of the Methods made use of in Laying the Foundations of the Piers of Westminster-Bridge,* London 1739 (copy illustrated by Gayfere in ICE library).

1743  *The Present State of Westminster Bridge . . . in a letter to a Friend* (containing a description of the bridge on 8 December 1742, probably by Labelye), London 1743.

1744  Prospectus of Labelye's proposed book, published as an Appendix in his *Description of Westminster Bridge,* 1751.

1745  *The Result of a View of the Great Level of the Fens taken in July 1745,* London 1745.

1746  Labelye's suggestions for London Bridge, see Maitland, *History of London,* 1746, II, p 829.

1747  *The Result of a View and Survey of Yarmouth Harbour taken in the Year 1747,* Norwich 1747.

1748  *Report relating to the Improvement of the River Wear and Port of Sunderland,* London 1748.

1748  *The Result of a Particular View of the North Level of the Fens. Taken in August 1745,* London 1748.

1751  *A Description of Westminster Bridge . . with an Appendix,* London 1751 (Telford's copy in ICE library).


Langley, Batty

*A Design for the Bridge at New Palace Yard, Westminster,* London 1736.
*A Reply to Mr J. James's Review of the Several Pamphlets & Schemes . . .* 1737.
*A Survey of Westminster Bridge as 'tis now Sinking into Ruin,* London 1748.
*Observations on a Pamphlet* (see Marquand).


Lediard, Thomas and Thomas Cotton

*A SCHEME humbly offered to the Hon the Commissioners for building a BRIDGE at Westminster for Opening Convenient and Advantageous Ways and Passages (on the West side) to and from the said Bridge, if situated at or near Palace Yard; as likewise to and from the Parliament House and Courts of Justice. With a Plan of Part of Westminster from the Hall to the Plantation Office. Thomas Lediard and Thomas Cotton, 15th February 1737–8.*


Lediard, Thomas

*A Letter from One of the Commissioners (signed N.N.) appointed by Act of Parliament for Building a Bridge at Westminster, to Mr Lediard, 25th February 1737–8. Mr Lediard's ANSWER, 12th March 1737–8.*

*Some Observations on the SCHEME offered by Messrs Cotton and Lediard, with a Plan of the lower parts of the Parishes of St Margaret and St John the Evangelist, 5th April 1738*. The BM copy (10349.g.37) is annotated by R. B. Prosser, author of the article on Labelye in DNB.

Marquand, Charles
*Remarks on the Different Construction of Bridges and Improvements to secure their Foundations,* 1749, together with Anon. (Batty Langley?) *Observations on a Pamphlet entitled 'Remarks on the Different Construction . . .'in which the Puerility of that Performance is considered,* 1749.

Price, John, *Some Considerations humbly offered to the Members of the House of Commons for Building a Stone-Bridge over the River Thames from Westminster to Lambeth, 25th February 1734–5*. Amended copy in Bodleian Library (Gough 22) contains a print showing three arches and four piers, one with a niche, another with œil-de- bœuf.

*REASONS against Building a Bridge over the Thames at Westminster,* undated broadsheet (1722?) in Bodleian Library (Gough 22).
*SCHEME for raising £60,000 for building a Bridge by means of a Toll,* undated (1736?) in Bodleian Library (Gough 22).

*Short Narrative of the Proceedings of the Gentlemen concerned in obtaining the Act for building a Bridge at Westminster (10th November 1737) with an M.P.'s Answer (24th November 1737), London 1737.*

BALLADS AND VERSES
Rusticus, 'On the Westminster-Bridge', *Gentleman's Magazine,* July 1736, p 417.
James Thompson, 'On the Report of a Wooden Bridge at Westminster', *Gentleman's Magazine,* August 1737, p 511, and *London Magazine,* September 1737, p 512.
Anon., 'The Downfall of Westminster Bridge or My Lord in the Suds', 1747 (BM 11626.h.11.7).
Anon., 'Westminster Bridge is sunk again', *London Magazine,* August 1748, p 374.
Anon., 'On a Nobleman [Pembroke] lately furnishing up the Back Front of his House by the Waterside, by a Waterman', *London Magazine,* 1749, p 572.
Anon., 'Who e'er this mighty frame surveys . . .', *London Magazine,* 1750, p 39.
Anon., 'On Westminster Bridge' (When late the river gods would visit Thames . . .), *London Magazine,* 1750, p 600, republished in *Gephyralogia,* 1751.
Anon., 'The Lament of Westminster Bridge', *Punch,* 1846, XI p 98.
Anon., 'The Downfall of Westminster Bridge', c 1850 (BM collection of broadsides, 11621.k.5.370).

NEWSPAPERS, JOURNALS AND MAGAZINES
British Museum, Burney Collection; Bodleian Library, Rawlinson Collection; London Library, Old Periodicals; Westminster City Reference Library.

*Annual Register*, 1758–
*Applebee's Original Weekly Journal*, 1720–36
*British Journal*, 1722–31
*Builder*, 1843–
*Champion*, 1739–43 (Fielding)
*Common Sense or the Englishman's Journal*, 1737–43
*Covent Garden Journal*, 1752 (Fielding)
*Craftsman*, 1726–50
*Daily Advertiser*, 1730–98
*Daily Courant*, 1702–35
*Daily Gazetteer and London Advertiser*, 1735–48
*Daily Post*, 1724–
*Fog's Weekly Journal*, 1716–37
*General Advertiser*, 1734–
*Gentleman's Magazine*, 1731–1907 (Dr Johnson and others)
*Historical Register*, 1716–38
*History of Parliament*, vols 7–14
*Illustrated London News*, 1842–
*London & Country Journal*, 1739–43
*London Chronicle*, 1757–1823
*London Daily Post*, 1730–40
*London Evening Post*, 1727–1806
*London Gazette*, 1666–
*London Gazetteer*, 1748–
*London Journal*, 1720–44
*London Magazine*, 1732–85
*Mist's Weekly Journal or Saturday Post* (later Fog's)
*New Miscellany for the Year*, 1734–9 (Swift and others)
*Political State of Great Britain*, 1711–37
*Post Boy*, 1728–36
*Public Advertiser*, 1734–94
*Punch*, 1841–
*Read's Weekly Journal or British Gazetteer*, 1715–61
*The Times*, 1785–
*Westminster Journal*, 1741–
*Westminster Magazine*, 1750–1
*Whitehall Evening Post*, 1718–38 and 1746–1801
*World*, 1753–6


GUIDE-BOOKS, LETTERS, DIARIES, ETC.
Macky, J., *A Journey through England* (1722–3).
Defoe, D., *A Tour through Great Britain*, 1724–7 (Everyman).
Anon., *A View of London and Westminster or the Town Spy* ... *by a German Gentleman* (second edition 1725).
Saussure, C. de, *A Foreign View of England in the Reigns of George I and George II*. 1725 (1902 edition).
Gonzalez, M., *Voyage to Great Britain*, 1730 (Pinkerton).

Hogarth, Scott, Thornhill, etc, *The Five Days' Peregrination,* 1732 (ed. Charles Mitchell, 1952).

Walpole, Horace, *Letters* (1732–97).

Anon., *Foreigners Guide to London* (1740).

Anon., *A Trip from St James's to the Royal Exchange with Remarks Serious and Diverting, on the Manners, Customs and Amusements of the Inhabitants of London and Westminster* (1744).

Kalm, Pehr, *A Visit to England* (1748).

Smollett, Tobias, *Travels through France and Italy* 1765 (Everyman).

Pococke, R., *Travels through England in 1750 and 1751,* (1889).

Hazlitt, W. C. (ed.), *Narrative of the Journey of an Irish Gentleman through England ... 1752,* (1869).

Angeloni, John, *Letters on the English Nation* (1756).

Grosley, P. J., *A Tour to London* (1756).

Russel, P. and Price, O., *England Displayed* (1769).

Archenholtz, J. W., *A Picture of England,* c 1780 (1797).

Moritz, Rev. C. P., *Journeys of a German in England in 1782* (ed. Reginald Nettel, 1965).

Hutton, William, *A Journey to London* (1874).

Wordsworth, Dorothy, *Journals* (1798–1828).

Geijer, Erik, *Impressions of England,* 1809–10 (1932).


BRIDGES I

Gautier, Hubert, *Traite des ponts* (1714)

Belidor, Forest de, *Architecture Hydraulique* (1737–53) *La Science de ingénieurs . . . d'architecture civile* (1729).

Anon., *Gephyralogia: an historical account of Bridges ancient and modern: including a particular history and description of the new Bridge at Westminster and an Abstract of the rules of Bridge-building* (1751).

Riou, Stephen, *Short Principles for the Architecture of Stone Bridges* (1760).

Hutton, Charles, *The Principles of Bridges* (1772), with a glossary.

Semple, George, *A Treatise on Building in Water* (1776).

Perronet, J. R. P. E., *Description des projets et de construction des ponts de Nouilly, de Nantes, d'Orléans et autres* (1782).


BRIDGES II

Gauthey, Navier, *Traité de la construction des ponts* (1809–22)

Le Sage, P. C., *Recueil des ponts* (1810).

*Report of the Middlesex Magistrates on Public Bridges* (1826).

Weale, John (ed.), *Theory, Practice and Architecture of Bridges* (1843).

London County Council, *Bridges, Tunnels and Ferries* (1898).

Dartein, F. de, *Etudes sur les ponts en pierre* (1907–12).

Webb, S. and B., *The Story of the King's Highway* (1913).

Home, Gordon, *Old London Bridge* (1931).

Kirby, R. S. and Lawson ,P. G., *The Early Years of Modern Civil Engineering* (Yale 1932).

Holdsworth, W. S., *History of English Law* X (1938), pp 322–32 for the legal responsibility of bridge maintenance.
Smith, H. S., *The World's Great Bridges* (1953).
De Maré, Eric, *Bridges of Britain* (1954).
Carsons, Patricia, 'The Building of the First Bridge at Westminster 1736–1750' in *Journal of Transport History* III (November 1957).


ROADS AND THE RIVER

Shapleigh, J., *Highways* (1749).
Brown, A., *Description of the Present Roads* ... (1765).
Hawkins, J., *Observations on the State of the Highways* (1763).
Homer, H., *Inquiry into the Means of Preserving and Improving the Public Roads* (1767).
Bayley, B., *Observations on the General Highway and Turnpike Acts* (1773).
Humpherus, Henry, *History* ... *of the Company of Watermen and Lightermen on the Thames* (1887).
Webb, S. and B., *The Story of the King's Highway* (1913).
Thacker, Frederick, *The Thames Highway*, 2 vols, 1914–20 (new edition 1968).
Jackman, W. T., *Development of Transportation in Modern England* (1916).
George, M. Dorothy, *London Life in the Eighteenth Century*, 1925 (new edition 1965) with a very full bibliography.
Brett-James, N. G., *The Growth of Stuart London*, 1935 (Ch. XVII, Traffic Problems).
Holdsworth, W. S., *History of English Law*, X (1938) pp 322–32 on legal responsibility for maintenance of bridges.
Phillips, Hugh, *The Thames about 1750* (1951).
Crackwell, B. E., *Portrait of London River* (1968).


PICTURES

Vertue, G., *Notebooks*, 6 vols with index in *Walpole Society Journals*.
Walpole, H., *Anecdotes of Painting in England*, 1762–71 (ed. Dallaway and Wornum, 1888).
*The Farington Diary*, 1793–1821 (ed. J. Greig, 1922–8); new edition to be edited by Kenneth Garlick.
Finberg, Hilda, 'Canaletto in England', *Walpole Society Journals*, IX and X, 1921–2.
Whitley, W. T., *Artists and their Friends in England, 1700–1799* (1928).
Finberg, H., 'Samuel Scott', *Burlington Magazine* (1942).
Parker, K. T., *Drawings of Canaletto at Windsor Castle* (1948).
Watson, F. J. B., *Canaletto* (1949 and 1954).
Messrs Thomas Agnew & Sons, Samuel Scott exhibition 1951 (introduction to *Catalogue* by Harald Peake).
Waterhouse, E. K., *Painting in Britain 1530 to 1790* (1953).
Moschini, V., *Canaletto* (1955).
Guildhall Art Gallery, Samuel Scott exhibition 1955 (introduction to *Catalogue* by J. L. Howgego).

Guildhall Art Gallery, William Marlow exhibition 1956 (introduction to *Catalogue* by J. L. Howgego).

Hayes, John, 'Parliament Street and Canaletto's View of Whitehall', *Burlington Magazine*, October 1958, pp 341–9.

Guildhall Art Gallery, *Canaletto in England* (1959).

Safarik, Eduard, *Canaletto's View of London* (Prague 1961).

Constable, W. G., *Canaletto,* 1963 (1976 edition revised by J. G. Links).

Millar, O., *The Tudor, Stuart and Early Georgian Pictures in the Collection of H.M. The Queen* (1963).

Holmes, M., 'London Scaffolding and a Brilliant Painting by Samuel Scott', *The Connoisseur,* June 1964.

Levey, M., *The Later Italian Pictures in the Collection of H.M. the Queen* (1964).

Hayes, John, 'A Panorama of the City and South London from Montague House by Robert Griffier' *Burlington Magazine* September 1965.

Guildhall Art Gallery, *Canaletto and his influence on London Artists,* 1965 (introduction to *Catalogue* by J. L. Howgego).

Millar, O., *The Later Georgian Pictures in the Collection of H.M. the Queen* (1969).

COLLECTIONS OF PRINTS AND DRAWINGS
Bodleian Library, Oxford
  Gough Collection, Westminster
  Sutherland Collection
British Museum, Department of Prints and Drawings
  Crace Collection
  Pennant's London, extra-illustrated editions
Department of the Environment
  General collection of prints, drawings and paintings
  Mason Beeton Collection
GLC Record Office, Gardner Collection and others
Guildhall
  Museum of London
  General Collection
  Pennant's London, extra-illustrated edition
Palace of Westminster
  Collection of prints and drawings relating to Westminster
Royal Institute of British Architects
  Collection of architectural prints and drawings
Sir John Soane's Museum
  Collection of architectural prints and drawings
Treasury
  Lister Collection, mainly of Whitehall
Victoria & Albert Museum, Department of Prints and Drawings
  General collection of topographical drawings
Westminster City Library, Buckingham Palace Road
  Gardner Collection and others relating to London and Westminster;
  Pennants London, extra-illustrated edition (A Westminster Bridge exhibition was held in the library in 1962).

Windsor Castle
Royal Collection of Canaletto drawings

GENERAL WORKS, EIGHTEENTH AND NINETEENTH CENTURIES
Bardwell, William, *Westminster Improvements* (1839).
(Bowles), *London Describ'd* (1731).
Chamberlain, H., *A Compleat History and Survey of London* (1770).
Concanen, Matthew, *History of Southwark* (1795).
Dodsley, R. & J., *London and Environs* (1761).
Entick, J., *A New and Accurate History and Survey of London* (1766).
Fielding, Sir John, *Inquiry into the Cause of the late Increase of Robbers* (1751).
Gwynn, John, *London and Westminster Improved* (1766).
Harrison, William, *A New and Universal History . . .* (1775).
Hone, W., *The Every-day Book . . .* (1826–7).
Jenner, Charles, *London Eclogues* (1773).
Lambert, B., *History and Survey of London and its Environs* (1806).
Lysons, Daniel, *Environs of London* (1792–6).
Maitland, William, *History and Survey of London . . .* (1739, 1756, 1772).
Malcolm, J. P., *Londinium Redivivum* (1802–7).
       ,,       *Anecdotes of the Manners and Customs of London . . .* (1808).
*Microcosm of London* (Ackermann), 1808–9.
Morris, Corbyn, *Observations on the Growth and Present State of London* (1751).
*New and Compleat Survey of London by a Citizen* (1742).
*New Remarks of London . . . by the Company of Parish Clerks* (1732).
Noorthouk, James, *History of London* (1773).
(Osborne), *English Architecture or the Public Buildings of London and Westminster* (1755).
Pennant, Thomas, *Some Account of London,* 1790 (extra-illustrated editions in British Museum Print Room, Westminster City Library, etc).
Ralph, J., *A Critical Review of Public Buildings . . . in and about London* (1734 and 1783).
Smith, J. T., *Antiquities of Westminster* (1807).
       ,,       *Additional Plates* (1809).
       ,,       *Nollekens and his Times,* 1829 (1920).
       ,,       *A Book for a Rainy Day,* 1845 (1905).
Stow, John, *Survey of London:* Strype's editions 1720 and 1754–5; Seymour's edition 1734–5.
Thornton, William, *History of London* (1784).
Walford and Thornbury, *Old and New London* (1876).

SOME LATER BOOKS
Colvin, H. M., *A Biographical Dictionary of English Architects 1660–1840* (1954), revised edition 1978.
Colvin, H. M. (ed.), *The History of the King's Works V 1660–1782* (1976) and *VI 1782–1851* (1973).
Darlington and Howgego, *Printed Maps of London circa 1553–1850* (1964).
De Maré, Eric, *London's Riverside* (1958).

George, M. D., *London Life in the Eighteenth Century* (1925), new edition 1966 with an invaluable bibliography.

London County Council, *Survey of London,* X, XIII and XIV, *St Margaret's Westminster* (1926–31); XXIII *Southbank and Vauxhall* (1951).

London Topographical Society publications, especially Porter's Map c 1660 in vol I, W. R. Lethaby Westminster in Vol VII, and Rocque's Plan 1746 in IX.

Phillips, Hugh, *Mid-Georgian London* (1964).

Plumb, J. H., *Sir Robert Walpole* (1956 and 1961), third vol forthcoming.

Rudé, George, *Hanoverian London 1714–1808* (1971).

Singer, Homyard, Hall & Williams, *A History of Technology* (1957) III.

Summerson, Sir John, *Georgian London* (1945), new edition 1970.

Summerson, Sir John, *Architecture in Britain 1530–1830* (1953), new edition 1963.

Trevelyan, G. M., *English Social History* (1944).

Turberville, A. S., *Johnson's England* (1933) with chapters on London by M. D. George and on architecture by Geoffrey Webb.

Victoria County History, *Middlesex* (1911–69).

Williams, Basil, *The Whig Supremacy 1714–60* (1939).

Withington, Robert, *English Pageantry* (1920), II pp 89–91 for the Lord Mayor's processions by barge to Westminster.

# Acknowledgements

So many of my friends have helped and encouraged me in writing this book on Westminster Bridge that I find it difficult to single them out. Will they please accept my thanks without being named. I would also like to thank the Master of the Rolls and the staff of the Public Record Office for allowing me special facilities for reading the many volumes of the Bridge Minutes in peace and quiet. This was a rare privilege which I greatly appreciate. Finally may I thank those private owners who have allowed me to illustrate their pictures and a bust without charge. In these commercial times this is an unusual generosity.

# Index